Complete Biology for Cambridge Secondary 1

Pam Large

Oxford excellence for Cambridge Secondary 1

OXFORD

Contents

Stage 7

1 Plants
- 1.1 Leaves, stems, and roots — 8
- 1.2 Enquiry: Questions, evidence, and explanations — 10
- 1.3 Review — 12

2 Humans
- 2.1 The human skeleton — 14
- 2.2 Muscles and movement — 16
- 2.3 Organ systems — 18
- 2.4 The circulatory system — 20
- 2.5 Studying the human body — 22
- 2.6 Extension: Extending lives — 24
- 2.7 Review — 26

3 Cells and organisms
- 3.1 The characteristics of living things — 28
- 3.2 Microbes — 30
- 3.3 Louis Pasteur — 32
- 3.4 Enquiry: Testing predictions — 34
- 3.5 Useful micro-organisms — 36
- 3.6 Enquiry: Planning investigations — 38
- 3.7 Harmful micro-organisms — 40
- 3.8 Plant and animal cells — 42
- 3.9 Specialised cells — 44
- 3.10 Extension: Nerves — 46
- 3.11 Tissues and organs — 48
- 3.12 Review — 50

4 Living things in their environment
- 4.1 Habitats — 52
- 4.2 Food chains — 54
- 4.3 Feeding ourselves — 56
- 4.4 Changing the planet — 58
- 4.5 Preventing extinction — 60
- 4.6 Obtaining energy — 62
- 4.7 Extension: Growing fuels — 64
- 4.8 Review — 66

5 Variation and classification
- 5.1 Variation — 68
- 5.2 Extension: Causes of variation — 70
- 5.3 Species — 72
- 5.4 Classification — 74
- 5.5 Vertebrates — 76
- 5.6 Classification of plants — 78
- 5.7 Review — 80

Stage 7 Review — 82

Stage 8

6 Plants
- 6.1 Why we need plants — 84
- 6.2 Enquiry: Asking scientific questions — 86
- 6.3 Water and minerals — 88
- 6.4 Review — 90

7 Diet
- 7.1 Food — 92
- 7.2 Enquiry: Managing variables — 94
- 7.3 A balanced diet — 96
- 7.4 Deficiencies — 98
- 7.5 Extension: Choosing foods — 100
- 7.6 Review — 102

8 Digestion
- 8.1 The digestive system — 104
- 8.2 Enzymes — 106
- 8.3 Extension: Using enzymes — 108
- 8.4 Review — 110

9 Circulation
- 9.1 Blood — 112
- 9.2 Extension: Anaemia — 114
- 9.3 The circulatory system — 116
- 9.4 Enquiry: Identifying trends — 118
- 9.5 Diet and fitness — 120
- 9.6 Review — 122

10 Respiration and breathing
- 10.1 Lungs — 124
- 10.2 Respiration and gas exchange — 126
- 10.3 Extension: Anaerobic respiration — 128
- 10.4 Smoking and lung damage — 130
- 10.5 Enquiry: Communicating findings — 132
- 10.6 Review — 134

11	**Reproduction and fetal development**	
11.1	Reproduction	136
11.2	Fetal development	138
11.3	Extension: Twins	140
11.4	Adolescence	142
11.5	Review	144

12	**Drugs and disease**	
12.1	Drugs	146
12.2	Disease	148
12.3	Extension: Defence against disease	150
12.4	Extension: Boosting your immunity	152
12.5	Review	154

Stage 8 Review   156

Stage 9

13	**Plants**	
13.1	Photosynthesis	158
13.2	Enquiry: Preliminary tests	160
13.3	Plant growth	162
13.4	Extension: Phytoextraction	164
13.5	Flowers	166
13.6	Seed dispersal	168
13.7	Review	170

14	**Adaptation and survival**	
14.1	Adaptation	172
14.2	Extreme adaptations	174
14.3	Extension: Survival	176
14.4	Enquiry: Sampling techniques	178
14.5	Studying the natural world	180
14.6	Review	182

15	**Energy flow**	
15.1	Food webs	184
15.2	Energy flow	186
15.3	Decomposers	188
15.4	Changing populations	190
15.5	Facing extinction	192
15.6	Extension: Maintaining biodiversity	194
15.7	Review	196

16	**Human influences**	
16.1	Air pollution	198
16.2	Enquiry: How scientists work	200
16.3	Water pollution	202
16.4	Saving rainforests	204
16.5	Review	206

17	**Variation and classification**	
17.1	Using keys	208
17.2	What makes us different?	210
17.3	Extension: Chromosomes	212
17.4	Extension: Investigating inheritance	214
17.5	Selective breeding	216
17.6	Enquiry: Developing a theory	218
17.7	Darwin's theory of evolution	220
17.8	Extension: Moving genes	222
17.9	Extension: Using genes	224
17.10	Review	226

Stage 9 Review   228

Reference pages   230
Glossary   242
Index   248

How to use your Student Book

Welcome to your **Complete Biology for Cambridge Secondary 1** Student Book. This book has been written to help you study Biology at all three stages of Cambridge Secondary 1.

Most of the pages in this book work like this:

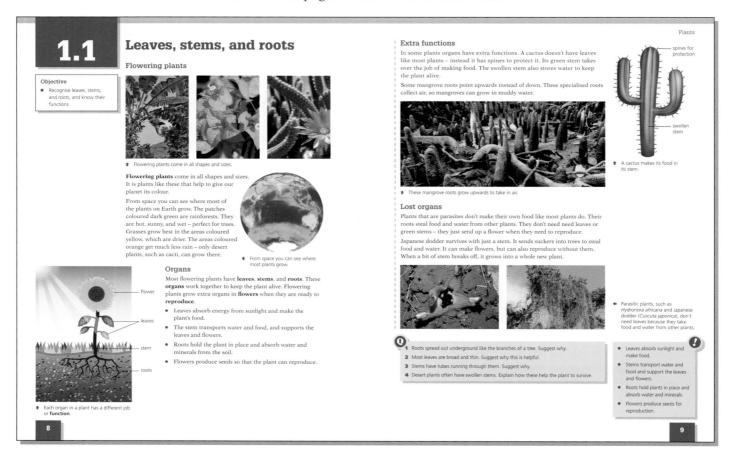

- Every page starts with the learning objectives for the lesson. The learning objectives are linked to the Cambridge Secondary 1 Science curriculum framework.
- New vocabulary is marked in bold. You can check the meaning of these words in the glossary at the back of the book.
- At the end of each page there are questions to test that you understand what you have learned.
- The key points to remember from the page are also summarised here.

These pages cover the Biology topics in the Cambridge Secondary 1 Science curriculum framework. In addition, in every chapter there are also pages that help you think like a scientist, prepare for the next level, and test your knowledge. Find out more about these below.

Scientific enquiry

These pages help you to practise the skills that you need to be a good scientist. They cover all the scientific enquiry learning objectives from the curriculum framework.

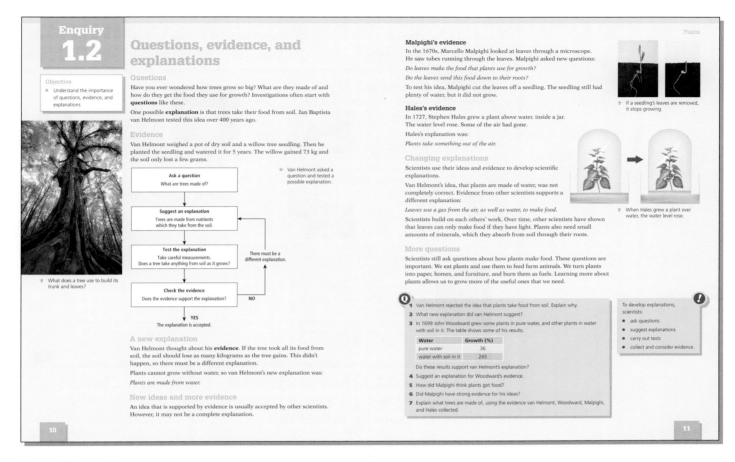

You will learn how to:

- consider ideas
- plan investigations and experiments
- record and analyse data
- evaluate evidence to draw scientific conclusions.

You will also learn how scientists throughout history and from around the globe created theories, carried out research, and drew conclusions about the world around them.

Extension

Throughout this book there are lots of opportunities to learn even more about biology beyond the Cambridge Secondary 1 curriculum framework. These topics are called extension because they extend and develop your science skills even further.

You can tell when a topic is extension because it is marked with a dashed line, like the one on the left. Or when the page has a purple background, like below.

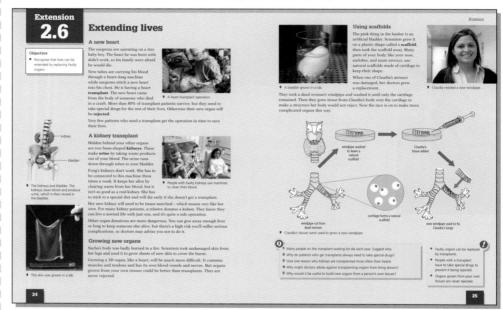

Extension topics will not be in your Cambridge Checkpoint test, but they will help you prepare for moving onto the next stage of the curriculum and eventually for Cambridge IGCSE® Biology.

Review

At the end of every chapter and every stage there are review questions.

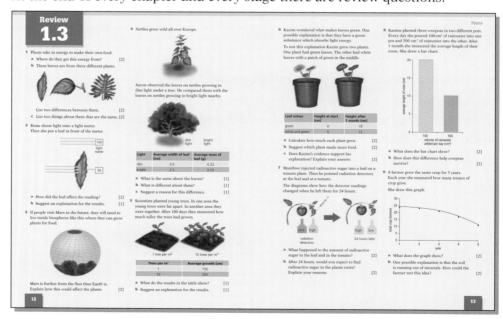

These questions are written in the style of Cambridge Checkpoint test. They are there to help you review what you have learned in that chapter or stage.

Reference

At the back of this book there are reference pages. These pages will be useful throughout every stage of Cambridge Secondary 1 Science.

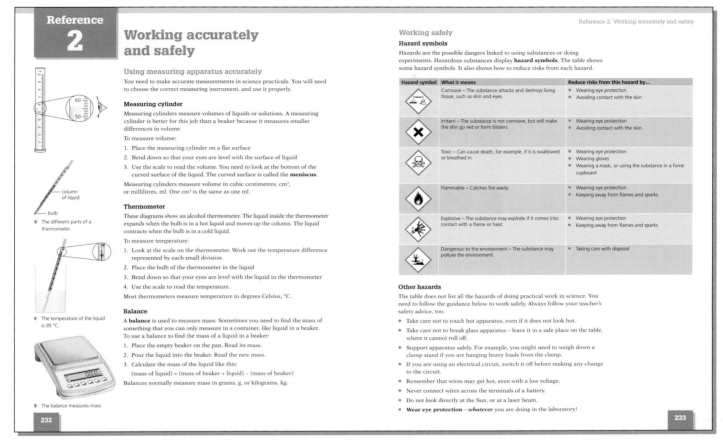

They include information on:

- how to choose suitable apparatus
- how to work accurately and safely
- how to detect gases
- how to record, display, and analyse results.

1.1 Leaves, stems, and roots

Flowering plants

Objective
- Recognise leaves, stems, and roots, and know their functions

↑ Flowering plants come in all shapes and sizes.

Flowering plants come in all shapes and sizes. It is plants like these that help to give our planet its colour.

From space you can see where most of the plants on Earth grow. The patches coloured dark green are rainforests. They are hot, sunny, and wet – perfect for trees. Grasses grow best in the areas coloured yellow, which are drier. The areas coloured orange get much less rain – only desert plants, such as cacti, can grow there.

↑ From space you can see where most plants grow.

Organs

Most flowering plants have **leaves**, **stems**, and **roots**. These **organs** work together to keep the plant alive. Flowering plants grow extra organs in **flowers** when they are ready to **reproduce**.

- Leaves absorb energy from sunlight and make the plant's food.
- The stem transports water and food, and supports the leaves and flowers.
- Roots hold the plant in place and absorb water and minerals from the soil.
- Flowers produce seeds so that the plant can reproduce.

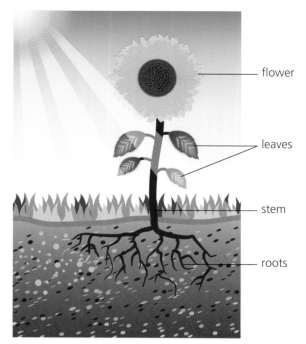

↑ Each organ in a plant has a different job or **function**.

Extra functions

In some plants organs have extra functions. A cactus doesn't have leaves like most plants – instead it has spines to protect it. Its green stem takes over the job of making food. The swollen stem also stores water to keep the plant alive.

Some mangrove roots point upwards instead of down. These specialised roots collect air, so mangroves can grow in muddy water.

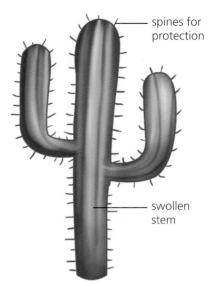

↑ A cactus makes its food in its stem.

↑ These mangrove roots grow upwards to take in air.

Lost organs

Plants that are parasites don't make their own food like most plants do. Their roots steal food and water from other plants. They don't need need leaves or green stems – they just send up a flower when they need to reproduce.

Japanese dodder survives with just a stem. It sends suckers into trees to steal food and water. It can make flowers, but can also reproduce without them. When a bit of stem breaks off, it grows into a whole new plant.

← Parasitic plants, such as *Hydronora africana* and Japanese dodder (*Cuscuta japonica*), don't need leaves because they take food and water from other plants.

Q
1. Roots spread out underground like the branches of a tree. Suggest why.
2. Most leaves are broad and thin. Suggest why this is helpful.
3. Stems have tubes running through them. Suggest why.
4. Desert plants often have swollen stems. Explain how these help the plant to survive.

- Leaves absorb sunlight and make food.
- Stems transport water and food and support the leaves and flowers.
- Roots hold plants in place and absorb water and minerals.
- Flowers produce seeds for reproduction.

Enquiry 1.2

Questions, evidence, and explanations

Objective
- Understand the importance of questions, evidence, and explanations

Questions

Have you ever wondered how trees grow so big? What are they made of and how do they get the food they use for growth? Investigations often start with **questions** like these.

One possible **explanation** is that trees take their food from soil. Jan Baptista van Helmont tested this idea over 400 years ago.

Evidence

Van Helmont weighed a pot of dry soil and a willow tree seedling. Then he planted the seedling and watered it for 5 years. The willow gained 73 kg and the soil only lost a few grams.

↑ What does a tree use to build its trunk and leaves?

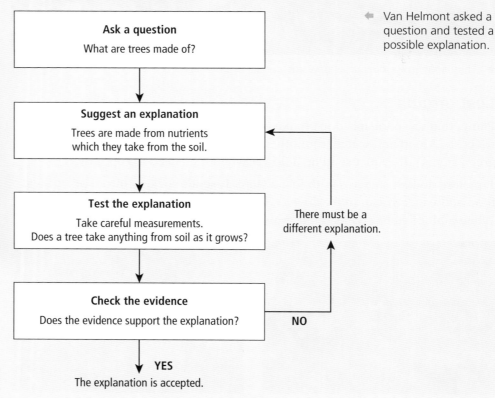

← Van Helmont asked a question and tested a possible explanation.

A new explanation

Van Helmont thought about his **evidence**. If the tree took all its food from soil, the soil should lose as many kilograms as the tree gains. This didn't happen, so there must be a different explanation.

Plants cannot grow without water, so van Helmont's new explanation was:

Plants are made from water.

New ideas and more evidence

An idea that is supported by evidence is usually accepted by other scientists. However, it may not be a complete explanation.

Malpighi's evidence

In the 1670s, Marcello Malpighi looked at leaves through a microscope. He saw tubes running through the leaves. Malpighi asked new questions:

Do leaves make the food that plants use for growth?

Do the leaves send this food down to their roots?

To test his idea, Malpighi cut the leaves off a seedling. The seedling still had plenty of water, but it did not grow.

Hales's evidence

In 1727, Stephen Hales grew a plant above water, inside a jar. The water level rose. Some of the air had gone.

Hales's explanation was:

Plants take something out of the air.

⬆ If a seedling's leaves are removed, it stops growing.

Changing explanations

Scientists use their ideas and evidence to develop scientific explanations.

Van Helmont's idea, that plants are made of water, was not completely correct. Evidence from other scientists supports a different explanation:

Leaves use a gas from the air, as well as water, to make food.

Scientists build on each others' work. Over time, other scientists have shown that leaves can only make food if they have light. Plants also need small amounts of minerals, which they absorb from soil through their roots.

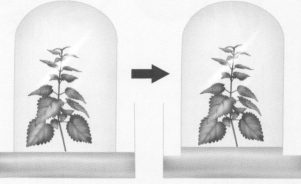

⬆ When Hales grew a plant over water, the water level rose.

More questions

Scientists still ask questions about how plants make food. These questions are important. We eat plants and use them to feed farm animals. We turn plants into paper, homes, and furniture, and burn them as fuels. Learning more about plants allows us to grow more of the useful ones that we need.

Q

1. Van Helmont rejected the idea that plants take food from soil. Explain why.
2. What was van Helmont's new explanation?
3. In 1699 John Woodward grew some plants in pure water, and other plants in water with soil in it. The table shows some of his results.

Water	Growth (%)
pure water	36
water with soil in it	265

   Do these results support van Helmont's explanation?
4. Suggest an explanation for Woodward's evidence.
5. How did Malpighi think plants got food?
6. Did Malpighi have strong evidence for his ideas?
7. Explain what trees are made of, using the evidence van Helmont, Woodward, Malpighi, and Hales collected.

To develop explanations, scientists:
- ask questions
- suggest explanations
- carry out tests
- collect and consider evidence.

Review 1.3

1. Plants take in energy to make their own food.
   a. Where do they get this energy from? [2]
   b. These leaves are from three different plants.

   List two differences between them. [2]
   c. List two things about them that are the same. [2]

2. Rima shone light onto a light meter.
   Then she put a leaf in front of the meter.

   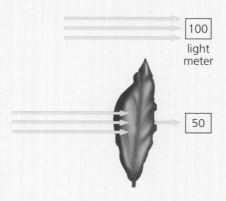

   a. How did the leaf affect the reading? [2]
   b. Suggest an explanation for the results. [1]

3. If people visit Mars in the future, they will need to live inside biospheres like this where they can grow plants for food.

   Mars is further from the Sun than Earth is. Explain how this could affect the plants. [2]

4. Nettles grow wild all over Europe.

   Aaron observed the leaves on nettles growing in dim light under a tree. He compared them with the leaves on nettles growing in bright light nearby.

Light	Average width of leaf (cm)	Average mass of leaf (g)
dim	5.0	0.22
bright	2.5	0.22

   a. What is the same about the leaves? [1]
   b. What is different about them? [1]
   c. Suggest a reason for this difference. [1]

5. Scientists planted young trees. In one area the young trees were far apart. In another area they were together. After 100 days they measured how much taller the trees had grown.

Trees per m²	Average growth (cm)
1	150
10	200

   a. What do the results in the table show? [1]
   b. Suggest an explanation for the results. [1]

6 Kazim wondered what makes leaves green. One possible explanation is that they have a green substance which absorbs light energy.

To test this explanation Kazim grew two plants. One plant had green leaves. The other had white leaves with a patch of green in the middle.

Leaf colour	Height at start (cm)	Height after 1 month (cm)
green	6	18
white and green	6	12

a Calculate how much each plant grew. [2]

b Suggest which plant made more food. [2]

c Does Kazim's evidence support his explanation? Explain your answer. [2]

7 Mambwe injected radioactive sugar into a leaf on a tomato plant. Then he pointed radiation detectors at the leaf and at a tomato.

The diagrams show how the detector readings changed when he left them for 24 hours.

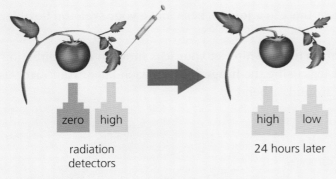

a What happened to the amount of radioactive sugar in the leaf and in the tomato? [2]

b After 24 hours, would you expect to find radioactive sugar in the plants' roots? Explain your reasons. [2]

8 Kanina planted three cowpeas in two different pots. Every day she poured 100 cm³ of rainwater into one pot and 500 cm³ of rainwater into the other. After 1 month she measured the average length of their roots. She drew a bar chart.

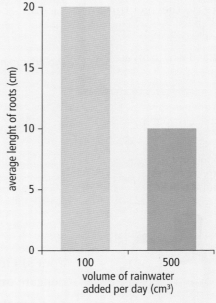

a What does the bar chart show? [2]

b How does this difference help cowpeas survive? [2]

9 A farmer grew the same crop for 5 years. Each year she measured how many tonnes of crop grew.

She drew this graph.

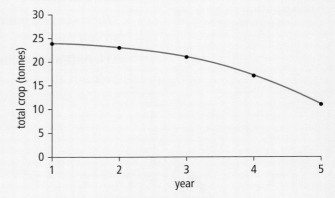

a What does the graph show? [2]

b One possible explanation is that the soil is running out of minerals. How could the farmer test this idea? [2]

2.1 The human skeleton

Support and protection

Objectives
- Describe the role of the skeleton
- Understand why joints are needed

↑ The column of vertebrae in your backbone form a bony cage around your spinal cord.

Your **skeleton** supports your body and helps you to move. It also protects delicate **organs** including your brain, heart, and lungs.

- Your **skull** surrounds your brain.
- A strong vertical column called your **backbone** holds up your head. This 'backbone' is made up of a stack of smaller bones called **vertebrae**. A thick bundle of nerves called your **spinal cord** runs down your back. It connects your brain to the rest of your body. The vertebrae form a strong cage around the cord to protect it from damage.
- A bigger cage, made of **rib** bones, supports and protects your heart and lungs.

Movement

Wherever two bones meet there is a **joint**. These make your skeleton flexible, so you can move.

The **ball-and-socket joints** in your hips let your legs move in every direction. You can also bend your legs using the **hinge joint** in each knee. Your arms use similar joints to swivel around or bend.

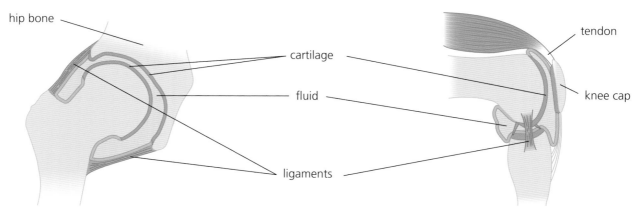

↑ Your hip joints are ball-and-socket joints.   ↑ Your knee joints are hinge joints.

Inside joints

Bones move smoothly against each other inside joints because joints are slippery inside. Two things make them like this.

- There is a smooth cushion of **cartilage** over the end of each bone.
- Most joints are full of a slippery liquid called **synovial fluid**.

Ligaments hold bones together but stretch just enough to let them move.

Injury

↑ Ball-and-socket joints at the hip and shoulder let our limbs move in every direction.

Footballers often injure their joints. As the ball shoots around the pitch they need to keep changing direction. Their sudden stops and starts can damage ligaments and cause a painful **sprain**.

If players collide, they may twist an arm bone out of its socket and **dislocate** their shoulder.

Even minor injuries increase a footballer's risk of developing **arthritis**. Arthritis can make synovial fluid disappear, cartilage wear away, and bones scrape together. Then any movement becomes very painful.

Q

1. List three reasons why you need a skeleton.
2. Explain how bones protect your brain, heart, and lungs.
3. Which of the labelled joints on the skeleton are hinge joints?
4. Which of the labelled joints on the skeleton are ball-and-socket joints?
5. Dancers create their steps using their knee and hip joints. In what way do knee joints differ from hip joints?
6. Draw a ball-and-socket joint. Add two labels to your drawing to show two things which help bones turn smoothly.
7. Choose a word from the list below to describe:

    a bone   b cartilage   c ligaments.

    flexible; hard; smooth

- A skeleton provides support and protection.
- The joints between bones let you move.

2.2 Muscles and movement

Muscles

Objectives
- Understand why muscles are arranged in pairs
- Predict what will happen when a given muscle contracts

Puppets move when strings pull on their arms and legs. Your own body moves when **muscles** pull on your bones.

Muscles can pull bones because they get shorter when they **contract**.

Muscles can only pull – they cannot push. So each muscle can only move a bone in one direction. Muscles work in pairs, with one muscle pulling a bone one way, and the other muscle pulling it back the other way.

⬆ Muscles work like a puppet's strings to pull our arms and legs around.

The biceps and triceps muscles in your arm work as a pair. When your biceps muscle contracts, your arm bends. At the same time, your triceps muscle relaxes and stretches. To straighten your arm, your triceps muscle contracts while the biceps relaxes and stretches.

The biceps and triceps are **antagonistic** muscles because they pull in opposite directions. Most movements are produced by antagonistic muscles.

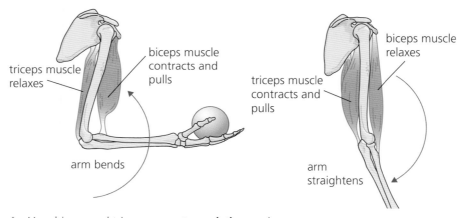

⬆ Your biceps and triceps are **antagonistic** muscles.

Muscles are attached to bones by strong cords called **tendons**. Tendons have to be very strong because muscles can exert enormous forces. Your Achilles tendons are the largest tendons in your body. They connect your calf muscles to your heel bones.

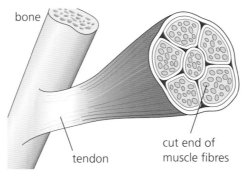

⬆ Tendons connect muscles to bones.

⬆ Trainers are designed to fit round your Achilles tendon.

Control of muscle movement

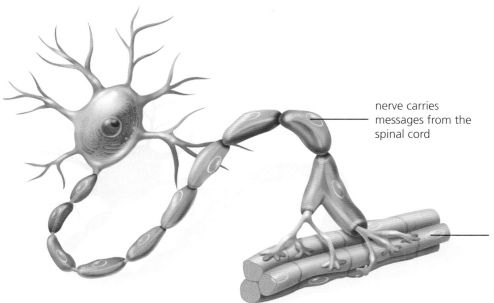

↑ Muscles are controlled by nerve messages.

Messages from your brain control when your muscles contract. These messages travel down through your spinal cord and along smaller nerves to every muscle fibre. If the backbone is damaged, this can stop these messages getting to muscles. Someone with a damaged spine may be paralysed.

Normally muscles take turns to contract, but an electric shock can make every muscle pull at once – suddenly and violently. This can jerk your bones and throw you across a room.

Strength

Muscles can only lift heavy loads for a short time. Then they need to rest and repair themselves. A robotic exoskeleton could give you extra strength. The messages your brain sends to muscles control the exoskeleton's motors as well. So it copies every move you make, but with 20 times the strength.

Exoskeletons can even help people with damaged spines to walk again.

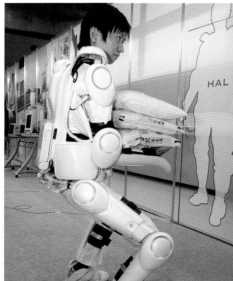

↑ A robotic exoskeleton could make you stronger.

Q
1. How do muscles cause movement?
2. What are antagonistic muscles?
3. Balance a book on the palm of your hand and hold it in front of you. Which muscle feels firmer, the one at the front of your upper arm or the one at the back? Explain why.
4. Antagonistic muscles also make your legs bend and straighten. Which muscle bends your leg – the one at the front, or the one at the back?
5. Feel the inside of your elbow as you lift a book. You should feel a strong cord. What is this? What does it do?
6. What tells a muscle when to contract?

- Muscles pull on bones to make you move.
- Antagonistic muscles pull bones in opposite directions.

2.3 Organ systems

Objectives
- Recognise the nervous, digestive, and respiratory systems
- Describe what these organ systems do

Organs

Your body is packed with different **organs**. Your brain, heart, lungs, liver, stomach, and intestines are all organs.

If your brain stopped working, you would be paralysed and unconscious. If your heart stopped, your blood would not go round your body. Each organ is **specialised**, which means it does one job very well. Different organs have different functions, and they work together to keep you alive.

Your skin is your biggest organ. It covers and protects all the others.

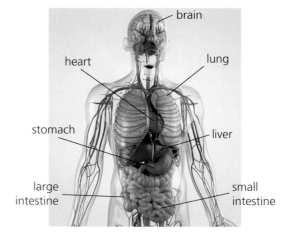

↑ Your brain, heart, lungs, liver, stomach and intestines are all organs.

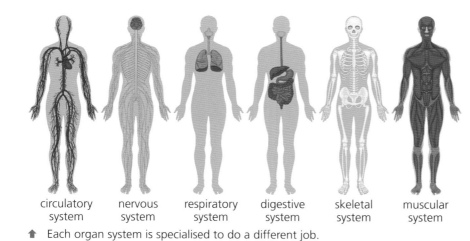

↑ Each organ system is specialised to do a different job.

circulatory system | nervous system | respiratory system | digestive system | skeletal system | muscular system

Systems

Groups of organs that work together form **systems**. The biggest systems are:

- the **skeletal system**, which contains all the bones in your skeleton
- the **muscular system**, all the muscles that move the bones.

The systems all work together and rely on each other. To move your bones, the muscles need a **nervous system** to control them and a **circulatory system** to bring them glucose and oxygen. They also rely on the **respiratory system** to keep the blood full of oxygen and the **digestive system** to take in glucose.

Nervous system

The **nervous system** controls how you respond to things around you.

When a friend enters the room, your eyes detect the light that reflects off his face. They send messages along nerves to your brain. In a fraction of a second, your brain works out what they mean. It sends new messages to muscles in your arm, mouth, and neck. You wave and say 'Hi' as you realise your friend has arrived.

The nervous system is made up of organs including the brain, spinal cord, nerves, and **sense organs** like your eyes and ears. These sense organs make you aware of what is happening around you.

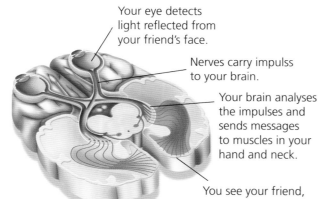

- Your eye detects light reflected from your friend's face.
- Nerves carry impulss to your brain.
- Your brain analyses the impulses and sends messages to muscles in your hand and neck.
- You see your friend, wave and say Hi!

↑ This model shows a slice through the brain showing how nerves link the brain to the eyes.

Respiratory system

Respiration is an important chemical reaction in your body, which releases the energy you need to stay alive. The reaction needs oxygen, and it makes a waste product called carbon dioxide. Your body needs to take in oxygen, and remove carbon dioxide, all the time.

Are you sitting still? If so, you are probably taking about 15 breaths per minute. Each breath brings fresh oxygen into your lungs. Your lungs are part of the respiratory system.

As blood rushes through your lungs, it swaps or exchanges the waste carbon dioxide gas for oxygen gas. So the function your respiratory system carries out is called **gas exchange**. To make gas exchange happen quickly, the air and blood are as close together as possible in the lungs.

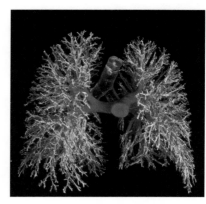

⬆ The white tubes carry air and the red tubes carry blood.

Digestive system

Energy drinks can give you a quick burst of energy. They contain sugars like **glucose**. During respiration, the glucose reacts with oxygen.

Glucose particles are tiny. As soon as you swallow them they pass into the blood vessels around your **small intestine**.

Other foods such as **starch** are made from bigger particles. The particles need to be broken down before they can get into your blood. Breaking down large particles is called **digestion**, and it happens in the **digestive system**.

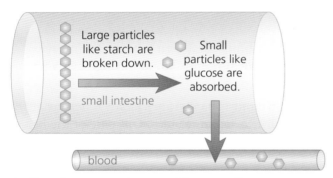

⬆ Digestion converts large particles to smaller ones.

Q

1 Which organ system controls how you respond to your surroundings?
2 You bite into a fruit. It tastes bitter. It could be poisonous. You spit it out. What has happened in your nervous system to make you do this?
3 Which organ system takes in nutrients like glucose?
4 Why does your body need oxygen?
5 What happens to your blood as it passes through your lungs?
6 What does your digestive system do to food?

- Your nervous system senses your surroundings and controls your body.
- Your respiratory system carries out gas exchange. Here oxygen moves into your blood and carbon dioxide moves out.
- Your digestive system breaks down food particles to allow them to get into your blood.

2.4 The circulatory system

Changing ideas

Your heart, blood, and blood vessels make up your circulatory system.

There have been several ideas about how the circulatory system works.

Around 2000 years ago, a Greek philosopher called Galen said that the heart made blood and pumped it to every part of the body, where it was used up. This explanation was accepted in Europe until the 1600s.

Then a doctor called William Harvey watched animal hearts beating. They pumped out massive amounts of blood. He realised new blood could never be made that fast. His new explanation was that the same blood must keep going round and round.

Years later Syrian manuscripts were translated and Europeans realised that an Islamic doctor, Ibn al-Nafis, had written more or less the same thing 350 years before Harvey's discoveries.

Objective
- Recognise the circulatory system and describe what it does

↑ Ibn al-Nafis explained how blood circulated.

The heart

Your **heart** is made of muscle. When it beats, the muscle contracts. Each heartbeat pumps blood through every part of your body.

Your heart beats around 100 000 times a day. It acts as a double pump, with two separate sides. One side of the heart sends blood to your lungs. The other sends blood to all the other parts of your body.

The blood leaves the heart through **arteries** and returns through **veins**. Between each artery and vein is a network of tubes called **capillaries** which have very thin walls. These let substances move in and out of your blood.

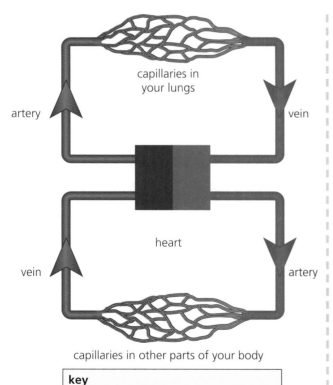

↑ Your heart pumps blood through your lungs and to other parts of your body.

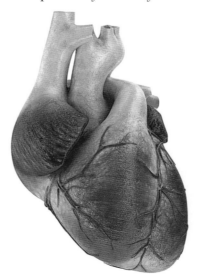

↑ Your heart is made of muscle.

Bringing supplies

Your body needs energy to stay alive. Energy can't be stored. You release it when you need it using the chemical reaction called respiration.

Respiration needs glucose and oxygen.

Blood collects glucose from your small intestine and oxygen from your lungs. It delivers these essential supplies to every part of your body, so they can all carry out respiration to supply the energy they need.

glucose + oxygen →(respiration) carbon dioxide + water
energy

↑ Your body gets energy by reacting glucose with oxygen.

Heart attack

Sam's heart has stopped beating. He's having a heart attack. The doctors are trying to save him. One squeezes air into his lungs. Another presses and releases his chest. This squeezes his heart, pushing blood around his body until they can start his heart beating again.

Heart attacks cause 1 in 8 deaths worldwide, but they are more common in some countries than others.

Sam's heart attack started when some of his heart muscles stopped working. The artery that brought blood to the heart muscle was blocked. The heart muscle couldn't get glucose or oxygen, so it ran out of energy and died.

Sam was lucky. His heart restarted. It wasn't too badly damaged. About half of all the people who have a heart attack live for another 10 years.

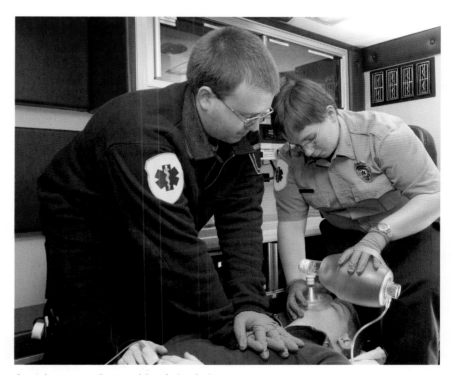

↑ A heart attack stops blood circulating.

Q
1. Harvey didn't believe Galen's idea that the heart made blood. Why not?
2. Suggest why the discoveries made by Ibn al-Nafis were not known in Europe in the 1600s.
3. Name three types of blood vessel.
4. Some blood has just left your lungs. Describe the journey it takes to get back to your lungs.
5. Describe one difference between your arteries and veins.
6. Explain why people can die within minutes if their heart stops beating.

- Your heart pumps blood through your lungs and around the rest of your body.
- Blood delivers glucose and oxygen to every part of your body.
- Blood leaves your heart in arteries and returns in veins.

2.5 Studying the human body

Objective
- Recognise that many scientists study the human body

Sports scientist

Hasan is wearing a mask as he runs. A sports scientist is measuring how much oxygen he takes in, and how fast his heart is beating. This data will show how fit he is and how successful his training has been.

Sports scientists also study the way athletes move. They can show athletes how to improve their performance and advise them how to avoid injury.

Neuroscientist

Nadia is a neuroscientist. She is testing a medicine for people with brain disease.

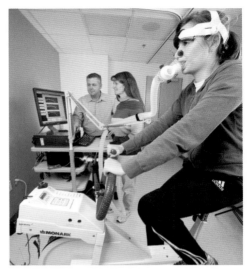
⬆ A fitness test shows how well your heart and lungs work.

Neuroscientists investigate how our brains and nervous systems work. They try to understand why we behave the way we do, and what causes mental illness. They also produce medicines for patients with problems like memory loss and depression.

➡ This scientist is testing a new medicine.

⬇ A prosthetic hand.

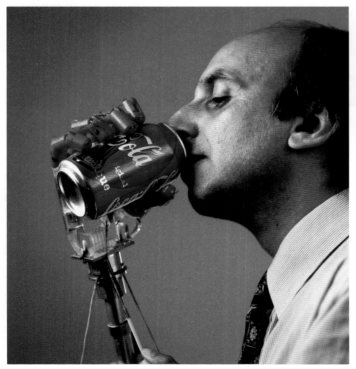

Prosthetic limb developer

A prosthesis is a device that replaces a missing body part. Dr Rahnejat developed this prosthetic hand for people who have lost their hand in an accident. It is designed to be as much like a real hand as possible. Its metal bones and tendons will be covered by a natural-looking skin.

If prosthetic hands can be connected to real nerves, patients can control them automatically and even feel what they are touching.

Haematologist

Many diseases make you feel tired or unwell. If a doctor doesn't know what's wrong, they will send a sample of your blood to a haematologist.

Haematologists study the way your blood changes when you are ill. Their test results can show doctors exactly what the problem is.

Dietician

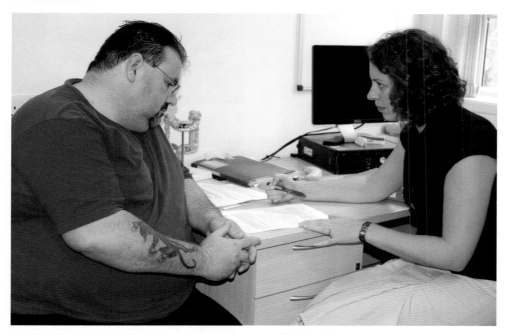

↑ Dieticians show patients what to eat to improve their health.

Dieticians study the connection between what you eat and your health. They use their knowledge to advise patients how to improve their health and fitness. Many illnesses can be cured by eating the right combination of nutrients.

Optometrist

If you can't see clearly, an optometrist can test your eyes. They look for signs of disease and check how clearly you can see. They can improve most people's vision by giving them the right sort of glasses or contact lenses.

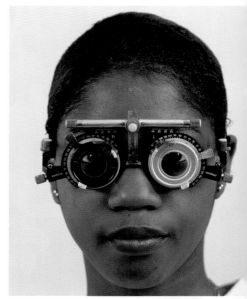

↑ Optometrists test your eyes.

Q
1. Sometimes special scientists help doctors to find out what is wrong with patients. Name one type of scientist and explain what they do?
2. Name a type of scientist who helps people recover from an illness. How do they do this?
3. Many other types of scientist study the human body. Choose two from the following list. Do some research and find out what they do: audiologist, cardiologist, dermatologist, and pathologist.

- Many different scientists study the human body.
- Their work helps us to stay healthy, recover from disease, or improve our lives.

Extension 2.6

Extending lives

A new heart

The surgeons are operating on a tiny baby boy. The heart he was born with didn't work, so his family were afraid he would die.

Now tubes are carrying his blood through a heart–lung machine while surgeons stitch a new heart into his chest. He is having a heart **transplant**. The new heart came from the body of someone who died in a crash. More than 80% of transplant patients survive, but they need to take special drugs for the rest of their lives. Otherwise their new organ will be **rejected**.

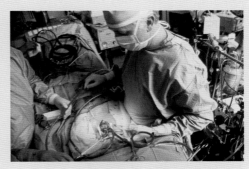

↑ A heart transplant operation.

Very few patients who need a transplant get the operation in time to save their lives.

A kidney transplant

Hidden behind your other organs are two bean-shaped **kidneys**. These make **urine** by taking waste products out of your blood. The urine runs down through tubes to your bladder.

Fong's kidneys don't work. She has to be connected to this machine three times a week. It keeps her alive by clearing waste from her blood, but it isn't as good as a real kidney. She has to stick to a special diet and will die early if she doesn't get a transplant.

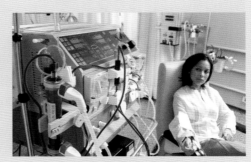

↑ People with faulty kidneys use machines to clean their blood.

Her new kidney will need to be tissue matched – which means very like her own. For many kidney patients, a relative donates a kidney. They know they can live a normal life with just one, and it's quite a safe operation.

Other organ donations are more dangerous. You can give away enough liver or lung to keep someone else alive, but there's a high risk you'll suffer serious complications, so doctors may advise you not to do it.

Growing new organs

Sacha's body was badly burned in a fire. Scientists took undamaged skin from her legs and used it to grow sheets of new skin to cover the burns.

Growing a 3D organ, like a heart, will be much more difficult. It contains muscles and tendons and has its own blood vessels and nerves. But organs grown from your own tissues could be better than transplants. They are never rejected.

Objective

- Recognise that lives can be extended by replacing faulty organs

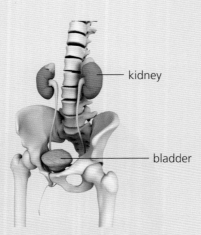

↑ The kidneys and bladder. The kidneys clean blood and produce urine, which is then stored in the bladder.

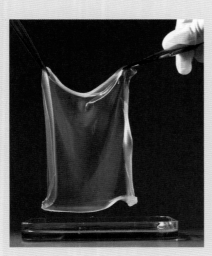

↑ This skin was grown in a lab.

Humans

Using scaffolds

The pink thing in the beaker is an artificial bladder. Scientists grew it on a plastic shape called a **scaffold**, then took the scaffold away. Many parts of your body, like your nose, earlobes, and main airways, use natural scaffolds made of cartilage to keep their shape.

When one of Claudia's airways was damaged, her doctors grew a replacement.

⬆ A bladder grown in a lab.

⬆ Claudia needed a new windpipe.

They took a dead woman's windpipe and washed it until only the cartilage remained. Then they grew tissue from Claudia's body over the cartilage to make a structure her body would not reject. Now the race is on to make more complicated organs this way.

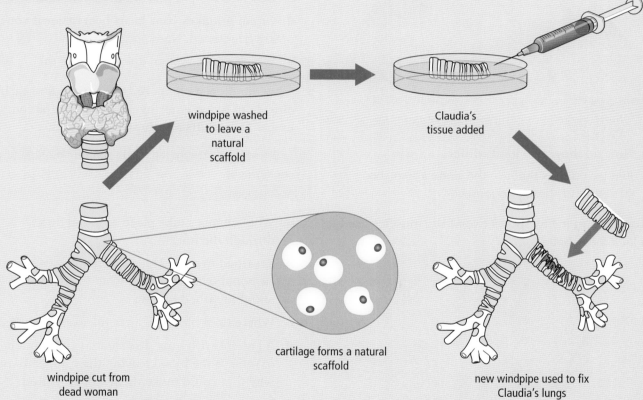

⬆ Claudia's tissues were used to grow a new windpipe.

Q

1. Many people die whilst they are waiting for an organ transplant. Why is this?
2. Why do patients who get transplants always need to take special drugs?
3. Why are kidneys transplanted more often than hearts.
4. Why might doctors advise against transplanting organs from living donors?
5. Why would it be useful to build new organs from a person's own tissues?

!

- Faulty organs can be replaced by transplants.
- People with a transplant have to take special drugs to prevent it being rejected.
- Organs grown from your own tissues are never rejected.

25

Review 2.7

1. Five patients have just arrived at the hospital.

   Karis: very pale, hardly breathing
   Amarjit: pale, no pulse
   Nadeen: back pain, can't move
   Ali: very weak, keeps being sick
   Simon: leg bone sticking out

   a Decide which organ system isn't working properly in each patient. [1]

   b Decide which two patients should be treated first and explain why. [2]

2.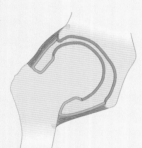

   a How do these two joints differ? [1]

   b List two things that are the same about the structure of these two joints. [2]

3. The diagram below shows two of the muscles used to kick a football.

   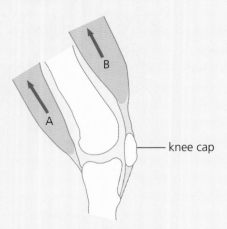

   a What happens to the footballer's leg when muscle A shortens? [1]

   b As he straightens his leg to kick, which muscle contracts and which relaxes? [2]

   c Explain why you need muscles on both sides of each leg bone. [1]

4.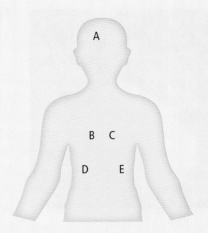

   a Which of the letters on the diagram shows the position of the heart? [1]

   b Name the organ system the heart belongs to. [1]

   c Describe the job this organ system does. [1]

   d Organ A makes your heart beat faster if you are frightened. Name organ A. [1]

   e Which organ system does A belong to? [1]

5. A haematologist compared the oxygen and carbon dioxide in blood entering and leaving a patient's lungs. The results are shown in the table.

Blood	Oxygen (units)	Carbon dioxide (units)
entering lungs	60	50
leaving lungs	100	40

   a What changes take place as blood flows through the lungs? [2]

   b Breathed-in air is 21% oxygen and 0.04% carbon dioxide. Suggest what is different about breathed-out air. [2]

   c Where does blood go after it leaves the lungs? [2]

   d A patient with lung disease may have much less oxygen in the blood leaving their lungs. Explain why that could make them feel very tired. [2]

6. Four students were asked about respiration.

   a Study their replies below and decide which student's answer is wrong. [1]

   Kevin: Respiration is what your lungs do.
   Sabrina: Respiration gives you energy.
   Tom: Respiration is a chemical reaction.
   Emmanuel: You need glucose and oxygen for respiration.

   b Explain the difference between respiration and gas exchange. [2]

7 When Jules climbed a mountain her hand got so cold that blood stopped circulating through her fingers.

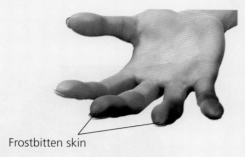
Frostbitten skin

Suggest why that made the ends of her fingers turn black and die. [2]

8 An average adult's heart beats between 70 and 80 times per minute. Regular exercise keeps a heart healthy. A sports scientist plotted a graph to show how much the heart rate should rise during exercise to keep the heart healthy.

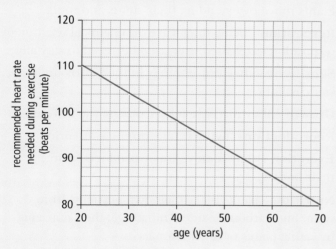

a What does the graph show? [1]

b What is the recommended heart rate during exercise for someone aged 50? [1]

c Suggest why the heart needs to pump faster during exercise. [2]

d Name the blood vessels that carry blood away from the heart. [1]

e Explain why the heart is called a double pump. [2]

f Which blood vessels return blood to the heart? [1]

9 A neuroscientist compared the time drivers took to put on their brakes when they spotted an accident. Her results are shown in the bar chart.

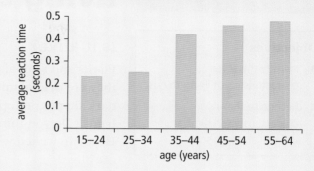

a What do the results show? [1]

b Explain how nerves and muscles can make a driver's foot move. [3]

10 A dietician measured the amount of glucose in an athlete's blood after they drank a glucose drink or ate a meal.

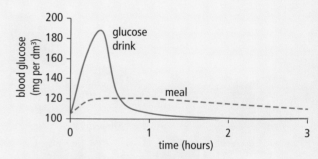

a What does the graph show? [3]

b What does the digestive system do? [2]

c Suggest why the glucose drink and the meal had different effects. [2]

11 Kasini collected data about broken bones in the lower body.

Part of the skeleton	Percentage of bones broken	
	Ages 10-19	Ages 60-69
hip	0	60
upper leg	6	21
knee	4	4
lower leg	20	5
ankle	25	10
foot	45	0

a What job do bones do? [2]

b Which bones are broken most often by each age group? [2]

c Which bones are broken 4 times more often by 10–19 year olds? [1]

3.1 The characteristics of living things

Objectives
- Identify the seven characteristics of living things
- Recognise these characteristics in familiar and unfamiliar organisms

Is it alive?

Plants and animals are living things – they are **organisms**. What about the dark shapes on top of this old tree trunk? Do you think they are alive? How can we tell whether things are alive?

Seven characteristics

We use the seven **characteristics** of living organisms to decide whether things are living or non-living. Living things all have these characteristics.

- **Movement** – most animals move from place to place. Plants move by extending their stems and roots.
- **Respiration** – living things use the chemical reaction of respiration to release the energy they need to stay alive.
- **Sensitivity** – they sense things in their surroundings, such as light.
- **Growth** – they increase in size during their lifetime.
- **Reproduction** – living things produce offspring.
- **Excretion** – they remove waste products from their bodies.
- **Nutrition** – plants make their own nutrients from water and carbon dioxide. Animals get their nutrients by eating plants or other animals.

Fungi are living things

The dark shapes in the top photo are part of a **fungus**. The rest of the fungus is a tangle of fine threads in the tree trunk underneath. As these threads grow, the fungus moves through the wood.

Fungi show all the characteristics of living things. They sense their surroundings, release energy using respiration, and excrete wastes like carbon dioxide. Fungi cannot make their own nutrients like plants, and they do not eat like animals. They absorb nutrients from their surroundings. This fungus dissolves wood to release the nutrients it needs.

Soon the dark shapes will release millions of tiny **spores** from which new fungi can grow.

Looking for life

Could there be life on Mars? The planet's surface looks lifeless, but perhaps there are living things hidden in the soil.

In 1976, scientists sent a probe to Mars to test this idea. The probe added nutrients to some Martian soil and warmed it up. Any life in the soil should use the nutrients for respiration, like life on Earth. That would produce wastes like carbon dioxide. Just as they hoped, some carbon dioxide formed.

A single piece of evidence can be misleading. Scientists like to have strong evidence from more than one experiment before they accept an explanation. The probe also tested the soil for chemicals found in living things. They found nothing. They were forced to accept a different explanation. There is nothing living in the Martian soil. The soil itself must be able to release carbon dioxide from nutrients. But could life have existed there in the past?

In 2012, a new probe called *Curiosity* began to collect evidence about Martian rocks. Rocks show scientists what the climate was like in the past. Perhaps there was a time when living things could exist on the planet's surface.

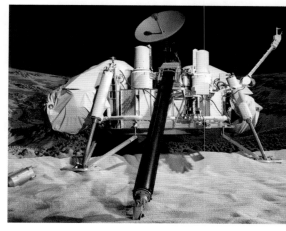

↑ This probe scooped up Martian soil to test it for signs of life.

Identifying life

Even here on Earth, life can be hard to detect. These grass seeds are dry. They don't show any of the characteristics of living things, but they are not dead. They can stay like this for thousands of years.

When they absorb water, they come to life. Each tiny seed uses nutrients stored inside it for respiration. It grows a tiny root and shoot. These sense which way to grow. The shoot extends to move into sunlight. Then it can make more nutrients so a whole new grass plant can grow.

↑ Dry grass seeds may not show any signs of life.

Q

1. Some people use the name 'Mrs Gren' to help them remember the seven things that all living things do. Match each letter in this name to one characteristic of living things.
2. Describe how plants and animals get their nutrients.
3. Most animals need to move around to survive. Most plants do not. What is the reason for this difference?
4. When seeds are soaked in water they begin to warm up and produce carbon dioxide. Explain why.

5. Look at the orange and blue shape in the picture above. Is it just a piece of coloured silk, or is it a living thing? Describe some tests you could do to collect evidence for your idea. Explain what results you would expect to get if you are right.

!

- All living things move, respire, sense, grow, reproduce, excrete, and use nutrients at some stage in their lives.

3.2 Microbes

Objectives
- Recognise four types of micro-organism
- Describe how a microscope works

Microscopic life

We are surrounded by living things that are too small to see. This idea was discussed from time to time in the last few centuries, but most people ignored it. Then, in the 1600s, **microscopes** were invented.

The idea could not be ignored any longer. Using microscopes, scientists saw tiny organisms wherever they looked. These ones were seen in a drop of pond water. Microscopic organisms are called **micro-organisms** or **microbes**.

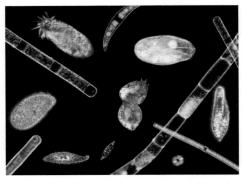

↑ Organisms in pond water – seen through a microscope

The first microscopes were very popular. One microscope maker wrote that people would 'cry out that they saw marvellous objects ... a new theatre of nature, another world'. Ideas about life itself were starting to change.

Microscopes

A school microscope uses light and lenses to **magnify** things. It can make them look up to 1000 times bigger. The **specimen** observed needs to be thin enough to let light pass through it.

In universities and research labs, scientists use **electron microscopes**. They are more complicated than light microscopes, but they give fantastic images – up to one million times the size of the specimen.

There are two sorts of electron microscope. SEMs (scanning electron microscopes) like this one show the surface of specimens. TEMs (transition electron microscopes) look through thin slices like light microscopes do. Images from electron microscopes are black and white but they can be coloured artificially to make them clearer.

You can change the **magnification** by using different **lenses**.

These knobs **focus** the image to make it clear and sharp.

A **mirror** reflects light up through the specimen.

Light passes through the thing you are looking at – the **specimen**.

↑ A light microscope

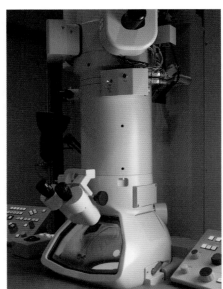

↑ A scanning electron microscope

Types of micro-organism
Algae and protozoa

There are two types of microbe in the photograph at the top of the page. The long green strands are **algae** and the others are **protozoa**.

It is easy to tell that the protozoa are alive because they move around in water. Algae make their own food like plants do, while protozoa feed on smaller micro-organisms, or take nutrients from their surroundings.

Fungi

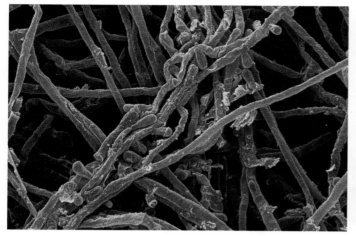

↑ An SEM image of a fungus from human skin – magnified around 2400 times.

↑ An SEM image of yeast cells from the skin of a grape – magnified 1930 times.

Most microscopic fungi are a tangle of thin threads called **hyphae**. They absorb nutrients from their surroundings. This fungus grows in human skin and nails. Similar fungi grow wherever there are nutrients.

Other fungi have no hyphae. They are called **yeasts**. They look like tiny spheres under the microscope. Some of them have smaller spheres stuck to their sides. They are reproducing. This type of reproduction is called **budding**.

Yeasts are found everywhere. You probably have a few living in your mouth. They are usually harmless.

Bacteria

Bacteria are the smallest micro-organisms. They are present wherever there are nutrients and water. They reproduce by splitting in two.

These bacteria are from someone's tongue. Other bacteria live in your digestive system. Most of the bacteria inside you are useful. They help you to digest your food.

↑ An SEM image of bacteria from someone's tongue – magnified 6700 times.

Viruses

The tiny green spheres around this bacterium are virus particles. **Viruses** cause diseases, but they are not alive. The only characteristic of living things they carry out is reproduction. They do this by entering living organisms and turning them into virus-making factories.

← A TEM image of virus particles surrounding a bacterium – magnified 48,000 times.

Q
1. Name four types of micro-organism.
2. Which type of micro-organism makes its own food, like plants do?
3. Which micro-organisms resemble tiny animals?
4. Which microbes are best seen through an electron microscope? Explain why.
5. Explain why viruses are not living things.

- Micro-organisms are living things that are too small to see without a microscope.
- They include algae, protozoa, fungi, and bacteria.

3.3 Louis Pasteur

Changing ideas

Objectives
- Recognise that micro-organisms can be useful
- Recognise the importance of Pasteur's studies

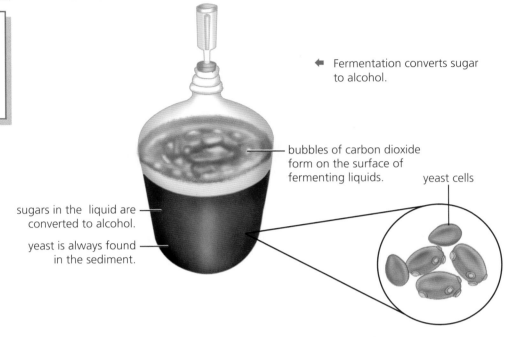

Fermentation converts sugar to alcohol.

bubbles of carbon dioxide form on the surface of fermenting liquids.

yeast cells

sugars in the liquid are converted to alcohol.

yeast is always found in the sediment.

Pasteur found out why wine and beer sometimes go bad.

Wine and beer are made by leaving sugary solutions to **ferment**. Bubbles form as they ferment and the sugar turns to alcohol.

Fermenting liquids always contain yeast. In the 1800s, scientists could see yeast under their microscopes, but they didn't realise these tiny spheres were alive.

In 1855, ideas about yeast began to change. Louis Pasteur was a professor of chemistry. He was studying wine and beer, which are made by fermentation. When he studied fermenting liquids he was convinced that yeasts were living things – tiny micro-organisms. He had a new idea – that yeasts use sugar for respiration and **alcohol** is one of their waste products.

More evidence

Pasteur was trying to find out why wine and beer sometimes go bad. He extended his idea. Sour milk contains **lactic acid**. If yeast makes alcohol, then different micro-organisms might make milk go sour by producing this lactic acid.

Just as he predicted, sour milk was full of micro-organisms – bacteria – which are even smaller than yeast. The bacteria use the sugar in milk for respiration and produce lactic acid as a waste. When Pasteur gave them extra sugar, their numbers grew. When he added them to milk, it went sour very quickly.

Beer that had gone bad always contained bacteria as well as yeast. Different bacteria caused different problems.

A problem solved

The wine and beer that went bad contained two sorts of micro-organism – yeast and bacteria. The drawing shows some of Pasteur's evidence. Beer could be spoiled in many different ways. For each type of spoilage, Pasteur found different bacteria mixed with the yeast. Each type of bacterium was adding different chemicals to the beer. They needed to be kept out or destroyed.

A new problem

Pasteur knew that bacteria could be destroyed by high temperatures, but that would also ruin the wine or beer. He tried keeping wine at 55 °C for a while. It was hot enough to kill most bacteria, but not so hot it spoiled the drinks. His idea worked.

Soon afterwards people all over the world were using Pasteur's method. They called it **pasteurisation**. Different temperatures and heating times were needed for each type of food or drink. Pasteurisation let them destroy most of the micro-organisms in foods and drinks without changing their flavours.

A microbe for every job

Pasteur realised that micro-organisms were responsible for many natural changes, like breaking down the bodies of dead plants and animals. Material made by living things is called **organic matter**. Micro-organisms help to break it down by using it as a source of nutrients.

When micro-organisms carry out respiration, some of the energy they release escapes as heat, so this pile of rotting plants is getting quite warm.

⬆ Most milk is pasteurised before it is sold.

⬅ Compost heaps get hot as micro-organisms break down the dead plant material.

Pasteur predicted that one day we would use micro-organisms to make many different chemicals. He was right.

Micro-organisms are now used to produce vast quantities of industrial chemicals, vitamins, and medicines. They are grown in large metal tanks called **fermenters**. Scientists separate out the useful products made by the micro-organisms from the water they have been grown in.

> **Q**
> 1 What is the name of the process that converts sugar to alcohol?
> 2 Explain why milk eventually goes sour.
> 3 Most milk is pasteurised before it is sold. Why does pasteurisation make milk last longer?
> 4 If a pile of dead leaves is left in a warm, damp place it gradually shrinks. Explain why.
> 5 What should be added to a fermenter to keep yeast growing inside it?

- Pasteur showed that yeast converts sugar to alcohol, and bacteria convert sugar to other products such as lactic acid.

Enquiry 3.4

Testing predictions

Objective
- Understand that scientists make predictions and check whether their evidence matches these predictions

Where do micro-organisms come from?

Pasteur showed that yeasts are living things, but he wondered where they came from? Why were they found in every fermenting liquid?

In the 1800s the idea of '**spontaneous generation**' was very popular. Scientists thought yeast was created every time a liquid fermented. Pasteur disagreed.

Pasteur's explanation was:

Yeasts are so tiny that they can blow around in the air. When they land in liquids full of nutrients they grow and reproduce, and this is what makes the liquid ferment.

Collecting evidence

Pasteur devised an experiment to test his explanation. He filled some flasks with a liquid full of nutrients. Then he stretched the neck of each flask into an S-shaped curve. Next he boiled the liquids to kill all the micro-organisms. So there were no micro-organisms present at the start, and none could get in from the air.

↑ Pasteur did not agree with the idea of spontaneous generation.

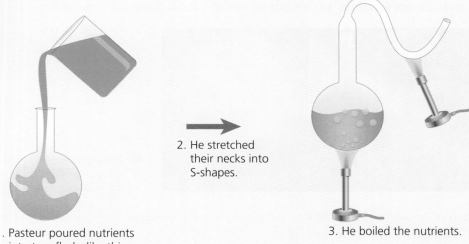

1. Pasteur poured nutrients into two flasks like this.
2. He stretched their necks into S-shapes.
3. He boiled the nutrients.

Making predictions

To test an explanation, scientists need to make **predictions**. They work out what results they will get if their ideas are correct.

Pasteur predicted that no micro-organisms would grow in his flasks because nothing could get into them. The liquid in the flasks should stay clear. But if he opened one flask, micro-organisms would get in. They would reproduce and turn the liquid cloudy.

Reviewing the evidence

Pasteur broke the neck of one flask so it was open to the air. The sealed flask stayed clear, and the open one went cloudy. The diagram on the next page shows his results.

Pasteur's results supported his explanation because they matched his prediction.

Pasteur had produced very strong evidence that spontaneous generation does not occur. Micro-organisms only form when their parents reproduce.

New ideas

Pasteur's work convinced him that infectious diseases could be spread through the air by micro-organisms. A Scottish surgeon called Joseph Lister heard about Pasteur's ideas. At the time, surgery was very dangerous. Half the patients who had operations became infected and died.

Lister predicted that more patients would survive if micro-organisms were kept out of their wounds. He found a chemical that would destroy micro-organisms – carbolic acid – and sprayed it over patients during their operations.

Lister's results matched his predictions and operations became much safer.

Eventually the sprays were abandoned. Instead surgeons began to wear gowns and masks, and the whole operating theatre was kept clean – not just the air above the patient.

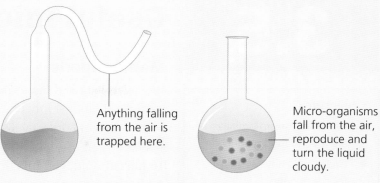

Anything falling from the air is trapped here.

Micro-organisms fall from the air, reproduce and turn the liquid cloudy.

⬆ Micro-organisms only grew in the flask with the broken neck.

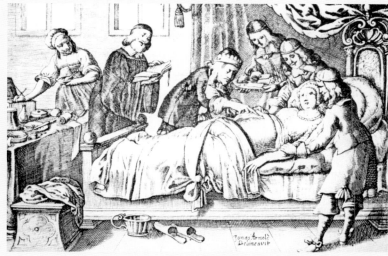

⬆ Lister's antiseptic spray killed any microorganisms around a patient's wound.

Q

1. When meat rots, it becomes infested with maggots. How would the theory of spontaneous generation explain this?
2. A scientist put fresh meat in two jars and left it to rot. One jar was open and the other was sealed. Flies flew in and out of the open jar and laid their eggs in the meat.
   a Predict which meat would fill with maggots if the theory of spontaneous generation was true?
   b Which meat would Pasteur expect to fill with maggots?
3. Look at Pasteur's flasks. Predict what would happen if a flask was tilted so that the nutrients ran into the S-shaped tube.
4. The skins of fruits such as grapes are normally covered with yeast. As an experiment, Pasteur compared two lots of grape juice. One lot was taken out of grapes with a syringe. The other was obtained by crushing grapes and draining out the juice. Predict which grape juice would ferment. Explain your reasons.
5. How could you stop a mixture of sugar, water, and yeast fermenting?
6. How could you tell whether a mixture of sugar and water had micro-organisms growing in it?

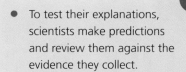

!
- To test their explanations, scientists make predictions and review them against the evidence they collect.

3.5 Useful micro-organisms

Objectives
- Know how micro-organisms are used in food production
- Understand how we make food last longer

Making foods

We use yeasts and bacteria to improve the flavour or texture of our foods, or to make them last longer.

Bread

To make bread, you mix flour with water to form **dough** which you then bake. Adding yeast to the dough before it is baked makes bread softer. The yeast uses some of the nutrients in the dough for respiration. This releases bubbles of carbon dioxide gas which make the bread soft and spongy.

The graph shows how yeast changed the volume of some dough when it was left in a warm place for an hour.

↑ Yeast is used to make bread rise.

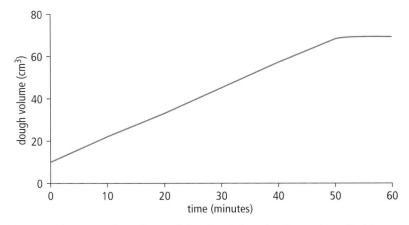

↑ Yeast increases the volume of the dough by releasing carbon dioxide.

Yoghurt and fermented milk drinks

Some bacteria get the energy they need by converting the sugar in milk to lactic acid. The acid gives yoghurt and fermented milk their flavour. It also makes these foods last longer by making them too acidic for other microbes to grow in it.

We can tell how acidic something is by measuring its pH. The lower the pH, the more acid it contains. The graph shows how the pH changes when bacteria ferment milk to make yoghurt.

↑ These pink *Lactobacillus* bacteria produce lactic acid

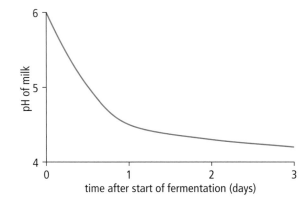

← The lower the pH, the more lactic acid the yoghurt contains.

Chocolate

It takes lots of different micro-organisms to make chocolate. To begin the process, cocoa pods are split open and the beans and pulp inside them are left to ferment. One after another, different fungi and bacteria feed on the pulp and release a mixture of chemicals that improve the flavour of the cocoa beans.

When the fermentation is complete, only the beans remain. They are dried, roasted, and powdered to make the cocoa that gives chocolate its flavour.

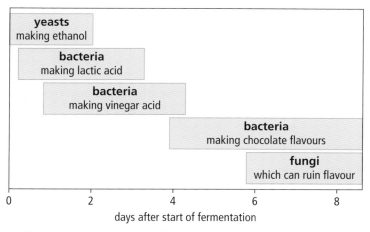

↑ These micro-organisms are involved in making chocolate.

↑ Chocolate is made from beans like these.

Keeping foods fresh

Some micro-organisms spoil food. We can slow down their growth by removing the oxygen, water, or warmth they need.

For long-term storage we can heat foods to **sterilise** them and then seal them in a plastic bag, jar, or can.

Some foods don't need to be kept cold and dry. They are too acidic or too salty for micro-organisms to grow in.

↑ When micro-organisms fall on food, they grow in it and make it rot.

Q

1. What type of micro-organism makes bread soft and spongy?
2. What type of micro-organism is used to make yoghurt?
3. According to the graph at the top of the opposite page, how does yeast change the dough volume when it is left in a warm place for 60 minutes?
4. The table below shows how temperature affects bread dough. Use the data to draw a graph. What does your graph show?

Temperature (°C)	Dough volume after 1 hour (cm³)
10	25
20	47
30	64
40	57
50	15

5. How does the pH of milk change when it is fermented to make yoghurt?
6. Suggest why it is important to ferment cocoa beans for the right length of time.

- Yeast produces carbon dioxide as it respires, so it can be used to make bread rise.
- Some bacteria produce lactic acid as they respire. They make cheese, yoghurt, and fermented milk drinks.

Enquiry 3.6

Planning investigations

Working with yeasts

Scientists keep finding new ways to use yeasts. Some yeasts make fuels we can use instead of petrol. Some make chemicals we can turn into plastics. Others can be used to clean up pollution.

For every new use, scientists ask an important question:

How can we make yeast grow and reproduce as fast as possible?

One possible idea is that yeast will grow faster if we keep it at a suitable temperature and provide all the nutrients it needs.

We can test this idea by doing an investigation. The temperature and types of nutrient are **variables**, which means things we can change or vary.

Objective
- Be able to plan an investigation

↑ Scientists investigate how well micro-organisms grow in different conditions.

Deciding what to change

To collect **valid** evidence we need to change one variable at a time and see whether it affects the yeast. We can compare our results more fairly if we collect accurate measurements.

Choosing what to measure

Five students planned to test the effect on yeast of changing the temperature. They all chose to take different measurements.

The best methods are quick and easy, and give accurate results. They should also give **repeatable** results, which means the results should be similar if they do the test again.

Controlled variables

All the students planned to measure how fast yeast makes carbon dioxide. This is a good way to show how fast it is growing. It makes carbon dioxide when it respires. The faster it grows and reproduces, the faster it respires.

Other variables, like the mass of yeast added at the start, could change the amount of carbon dioxide produced. So these variables need to be **controlled**. When the temperature is changed, everything else must be kept the same.

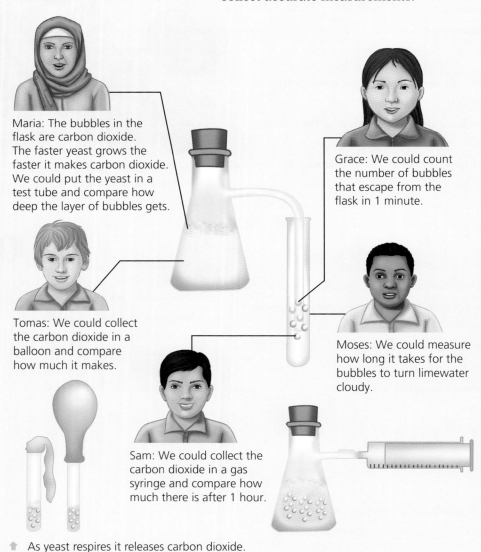

Maria: The bubbles in the flask are carbon dioxide. The faster yeast grows the faster it makes carbon dioxide. We could put the yeast in a test tube and compare how deep the layer of bubbles gets.

Tomas: We could collect the carbon dioxide in a balloon and compare how much it makes.

Sam: We could collect the carbon dioxide in a gas syringe and compare how much there is after 1 hour.

Grace: We could count the number of bubbles that escape from the flask in 1 minute.

Moses: We could measure how long it takes for the bubbles to turn limewater cloudy.

↑ As yeast respires it releases carbon dioxide.

38

Cells and organisms

Evidence

Grace's method did not work very well. The bubbles didn't come out in a steady flow.

Moses was not happy with his results either. At some temperatures the limewater did not go cloudy.

The other three students all finished their investigations. They got similar results.

When they test another variable they should do all their experiments at the best temperature.

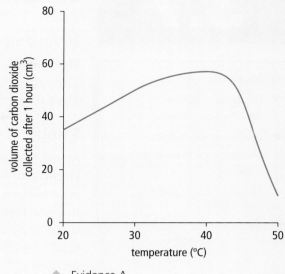

↑ Evidence A

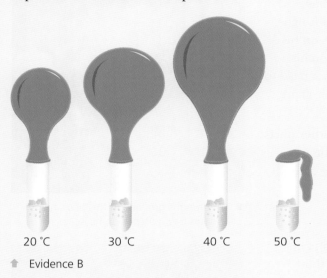

↑ Evidence B

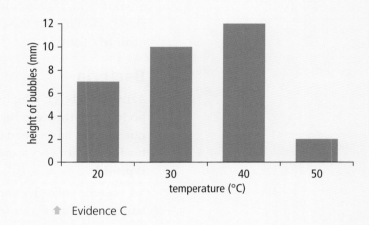

↑ Evidence C

Q

1. Why do scientists try to find the best conditions for growing each type of micro-organism?
2. Two variables that might affect the growth of yeast are the temperature and the types of nutrient present. Suggest another variable that might affect the yeast.
3. Choose one student's method. Draw a diagram to show their apparatus. Describe what measurements they need to take.
4. For the method you chose in question 3, list the variables the student would need to control.
5. Match each set of evidence to the correct student's method.
6. Estimate the temperature at which yeast grows fastest.
7. The same students want to test the effects different nutrients have on the same yeast. They plan to use the same method so they can compare their result more easily. Which method would you recommend they use? Explain why.

- Investigations usually start with a question.
- To plan an investigation:
  - decide what to change to answer the question
  - choose what to observe or measure to see the effects
  - identify any other variables that need to be controlled.

3.7 Harmful micro-organisms

Objectives

- Understand what is meant by an 'infectious disease'
- Give some examples of diseases caused by micro-organisms
- Suggest how to avoid infections

Invading microbes

Most micro-organisms are harmless, but some cause **infectious diseases**. They invade your body, use your tissues for food, and reproduce inside you. Then they find their way out to infect new victims.

Fungi

Ajani's head is very itchy. He has **ringworm**. This means that a fungus is growing in his skin. It feeds on his hair and leaves bald patches on his scalp. The fungus spreads from head to head by touch, or when people share brushes and combs.

Similar fungi can grow on other parts of your skin. If they infect the moist skin between your toes, they cause **athlete's foot**.

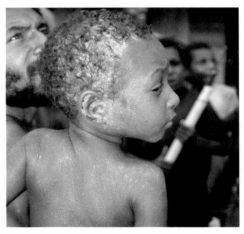

↑ The fungus growing in this boy's hair creates bald patches on his scalp.

Bacteria

Azizi has a fever, which means his body temperature is higher than normal. He has a headache and his **faeces** are very watery. He has **typhoid**.

You catch typhoid when harmful bacteria get into your food or drink. They reproduce in your digestive system and make you ill.

A typhoid victim's faeces are full of harmful bacteria. These can get on the person's hands and be passed to anything they touch. They can also get into drinking water, if the waste from toilets isn't cleaned up properly.

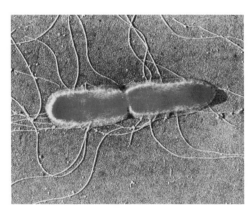

↑ Bacteria like these cause typhoid fever if they invade your digestive system – magnified over 2000 times.

Similar bacteria cause **food poisoning** which causes vomiting and diarrhoea. People can avoid passing on these bacteria by washing their hands after using the toilet.

Protozoa

Ayesha got bitten by a mosquito. It injected saliva into her blood and then sucked it up. The saliva contained protozoa, the parasites that cause **malaria**.

The protozoa will reproduce inside Ayesha's body. They will release chemicals that will give her a fever, chills, and muscle pains.

The mosquito is a **vector**. It carries protozoa from one victim to the next. Tsetse flies and sand flies also spread protozoa. Tsetse flies carry a fatal disease called **sleeping sickness** and sand flies carry **leishmaniasis**, which produces nasty sores on a victim's skin.

Viruses

Many viruses infect humans. The most common ones cause **colds** and **flu**. If you breathe in these virus particles they make you ill, but you usually get better quite quickly.

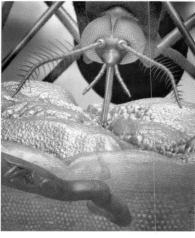

↑ The blue shapes in the blood vessel are the protozoa that cause malaria. A mosquito injects them as she prepares to suck blood.

normal liver   liver infected by hepatitis C

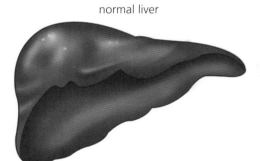 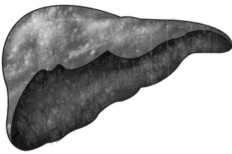

↑ The liver on the right has been damaged by hepatitis C.

Hepatitis C is different. If you get infected, the virus starts to damage your liver, but it can be years before you get ill. Eventually most victims begin to feel very tired. They may also get stomach pains or a fever. Many victims die because their liver stops working.

The virus can only be passed directly from one person's blood to another. This happens when more than one person has an injection from the same needle. It can also happen if you get a tattoo, or have your ears pierced, with contaminated equipment.

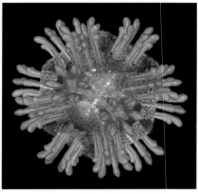

↑ Hepatitis C virus particles reproduce inside the liver and destroy it.

Q

1. Copy and complete this table to show two diseases caused by each type of micro-organism.

Micro-organisms	Diseases caused	How they get into the body
fungi		
bacteria		
protozoa		

2. Name two diseases caused by viruses and explain how you catch them.

3. Suppose you lived in an area where these diseases are common: ringworm, typhoid, malaria, and hepatitis C. Explain how you could avoid catching each disease.

- Infectious diseases are caused by micro-organisms.
- Different fungi, bacteria, protozoa, and viruses cause different diseases.

3.8 Plant and animal cells

Objectives
- Identify structures seen in plant and animal cells
- Compare plant and animal cells

Cells

The image on the right shows skin from the insides of your cheeks. It is magnified about 2000 times. Each of these tiny pink compartments is a **cell**.

Every part of your body is made of cells. So are all living things. Protozoa, yeasts, and bacteria only have one cell, but large plants and animals have billions.

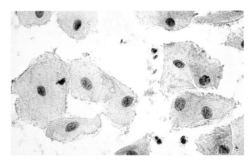

↑ Skin from the inside of your cheeks looks like this under the microscope.

The most important thing these cells do is stay alive. You are alive because your cells are all alive. They use energy from respiration, and simple nutrients, to build the spare parts they need to grow and repair themselves.

Animal cells

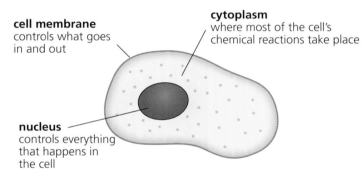

cell membrane controls what goes in and out

cytoplasm where most of the cell's chemical reactions take place

nucleus controls everything that happens in the cell

↑ A typical animal cell.

Cells come in many shapes and sizes. The diagram on the left shows a typical animal cell. It has three main parts with different functions.

- The **nucleus** stores a set of instructions called **genes**, which control what happens in the cell.
- Most of the chemical reactions that keep cells alive take place in the **cytoplasm**.
- The **cell membrane** controls what enters and leaves the cell. It lets nutrients and oxygen in and waste products out.

Plant cells

Like animal cells, plant cells have a nucleus, cell membrane, and cytoplasm. They also have extra parts so they can make their own food and support themselves.

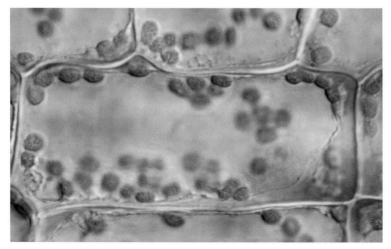

↑ This leaf cell has been magnified about 1000 times.

Cells and organisms

The green structures in most plant cells are **chloroplasts**. They capture light energy, which plants use to make food. The food is stored in cell sap, which fills a large bag called a **vacuole** in the middle of the cell. Root cells don't receive light, so they have no chloroplasts.

Every plant cell has a tough **cell wall** around its membrane. The vacuole pushes against the cell wall to keep a plant cell firm. Firm cells make a plant stand up straight.

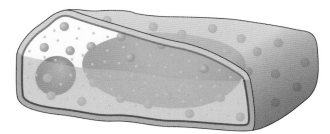

↑ A cells have a 3D shape like this cut-away plant cell.

Light microscopes can make cells look flat, but they aren't really. This cut-away diagram shows a plant cell's shape more clearly. We usually draw slices through cells to show what they contain.

Q

1. Draw a labelled diagram of an animal cell.
2. Copy this plant cell and put the correct labels on parts A to F.

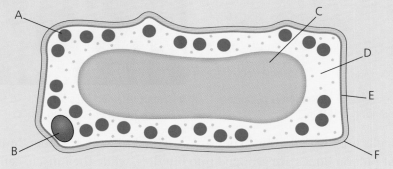

3. Leaves look green. Which part of a cell gives them their colour?
4. Copy this table. Complete the first two columns to show the parts of a cell and what they do.

Part of cell	Function	Plant, animal, both?

5. Complete the last column of your table to show whether each part is found in plants, animals, or both.
6. When a plant doesn't have enough water its leaves and stems bend downwards. This is known as wilting. Why does a plant wilt when it doesn't have enough water?

!

- All living things are made of cells.
- All animal and plant cells have a cell membrane, cytoplasm, and nucleus.
- Plant cells also have a cell wall and vacuole.
- Plants cells that grow in the light also have chloroplasts.

3.9 Specialised cells

Objective
- Relate the structure of cells to their functions

Cells with special jobs

In micro-organisms one cell has to carry out every life process. In larger organisms, cells are **specialised** for different functions. Each type of cell is **adapted** to carry out one life process really well.

Red blood cells

Blood looks like a liquid, but it's full of cells – mostly red blood cells. More than half the cells in your body are red blood cells. They are your smallest cells. These have been magnified about 2000 times. They are travelling through a blood vessel made from a tube of wider, flatter cells.

Red blood cells have no nucleus and they are small and flexible. They are packed full of **haemoglobin** – a red chemical that carries oxygen. They are specialised to squeeze through narrow blood vessels to deliver oxygen to every other cell.

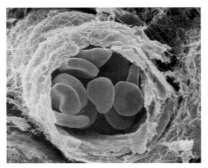

↑ Red blood cells carry oxygen to cells all over your body.

Muscle cells

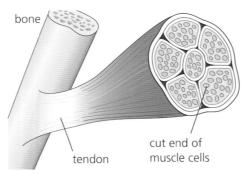

↑ Muscles cells are called fibres because they are long and thin.

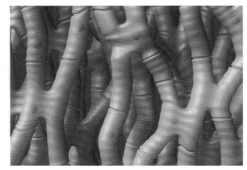

↑ Heart muscle fibres interlink so they can work together to pump blood.

Muscle cells are long and thin. They are designed to work together. Muscle cells that pull on bones are grouped into bundles, while the muscle cells in the heart are linked together.

Each muscle cell is completely full of **fibres**, so its nucleus is squeezed to the side. The fibres allow muscle cells to contract and produce movement.

Fat cells

Each of these big, pink cells is full of fat. The nucleus and cytoplasm are squeezed to the side to make room.

Fat cells act as an energy store for times when you can't eat enough. A layer of them under your skin also helps to keep you warm.

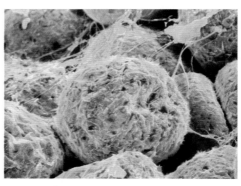

↑ Fat cells swell up to store excess food.

Bone cells

Bone is rigid and hard but it is full of living cells.

Bone cells make fibres and excrete them into their surroundings. Then minerals stick to these fibres and turn them into solid bone. The cells are left surrounded by bone. They reach out through tiny gaps to contact other cells and take in nutrients.

Specialised plant cells

Leaf cells absorb sunlight and make nutrients, while root cells are specialised to take in water and minerals.

The cells on the outside of this root are **root hair** cells. They have long, thin side branches. These increase the **surface area** of roots and make it easier for them to collect water and minerals from the soil around them.

The four green circles are **xylem** cells. They form tough hollow tubes that carry water from the roots to the rest of the plant. The blue **phloem** cells around them are specialised to carry nutrients down from the leaves.

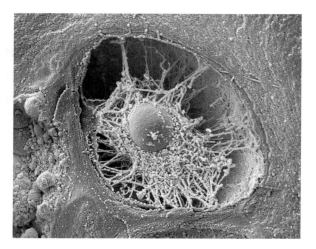

← A bone cell surrounded by bone.

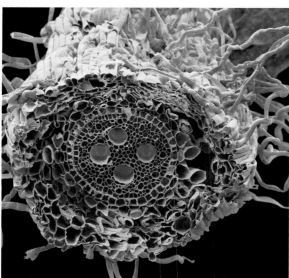

↑ Each type of root cell in this SEM image has been given a different colour.

Q

1. Draw each of these types of cell and explain why it is good at its job: a red blood cell; a muscle cell; a fat cell; a bone cell; and a plant's root hair cell.

2. Cells A, B and C are specialised cells. Draw each one and then suggest where you might find it.

   Cell A is a wide flat cell that only lives for a short time. It protects the cells underneath it.

   Cell B's cytoplasm forms lots of branches. It connects to many other cells to send and receive messages.

   Cell C has hundreds of tiny hairs on top of it so it looks like a brush. The hairs move to and fro to brush away any dirt that enters your body from the air.

- Plant and animal cells specialise to do different jobs.
- A red blood cell's adaptations make it good at carrying oxygen.
- Muscles cells specialise in making themselves shorter.
- Fat cells are adapted for storage and bone cells to build bones.
- Root hair cells specialise in taking in water and minerals.
- Xylem and phloem cells form tubes to transport water and nutrients.

Extension 3.10

Nerves

Objectives
- Relate the structure of nerve cells and sensory cells to their functions
- Recognise how explanations develop as new evidence becomes available

Knowledge about nerves

Scientists have known what nerves do for a long time. In 1025, Ibn Sina wrote one of the first medical textbooks. One of his ideas in the book was that nerves connect with the brain, make muscles move, and pick up pain messages.

He thought patients should be awake for some operations. When they screamed, that showed when the knife was cutting a live nerve – which proved all their dead tissues had been successfully removed.

↑ Ibn Sina.

Experiments with frogs

The first clue about how nerves sense pain and make muscles move came from experiments in Italy in the 1700s. Luigi Galvani noticed that frogs' legs twitched if he touched their nerves with a knife. Electrical sparks made them twitch even more.

After trying lots of different experiments, Galvani concluded:

Nerves carry electrical signals from the brain to muscles.

Examining the brain

When scientists started using microscopes they found cells in every organism. That raised questions. Could our brains be made of separate cells too? Could a collection of cells produce our thoughts and feelings?

↑ Luigi Galvani.

Eventually microscopes improved. Bundles of long thin cells were found inside nerves.

Then stains were developed that made the cells in thin slices of brain visible under the microscope.

By the 1890s, scientists had their evidence. Our brains really are made of separate nerve cells. There are billions of them, and they are all connected. On average, each nerve cell connects to 10 000 others.

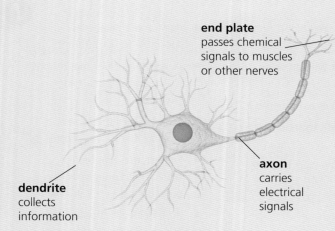

↑ Nerves contain bundles of nerve cell axons.

end plate passes chemical signals to muscles or other nerves

axon carries electrical signals

dendrite collects information

Cells and organisms

Another experiment with frogs

The next question scientists had to answer was:

How do nerves communicate with each other and give instructions to muscles?

One possible explanation was that they use chemicals to carry messages between cells.

In 1921, Otto Loewi tested his idea. He put beating frog's hearts in two separate containers. Then he sent an electrical signal down the nerve to one heart, to slow it down.

Next, he added some of the liquid around the first heart to the second heart. That slowed down too. There must have been a chemical in the liquid. He had strong evidence that nerves use chemicals to communicate.

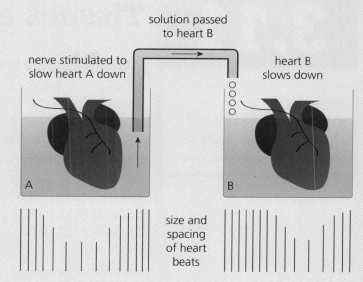

Sensing your surroundings

Sensory cells detect change. Everything you see, hear, taste, smell, or feel, and everything you know about your body – like what your arms and legs are doing – depends on the signals these cells send to your brain.

Sensory cells in your ears and eyes detect sound and light. But your ears don't hear and your eyes don't see. They just send electrical signals along nerves to your brain. The nerve cells in your brain turn these messages into sounds and vision.

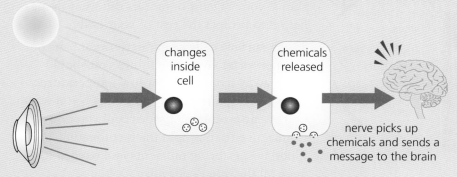

↑ Sensory cells detect change and send signals along nerves to the brain.

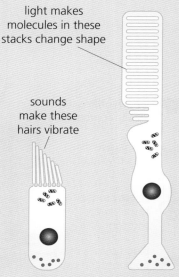

↑ Sensory cells detect change and send signals along nerves to the brain.

Sound- and light-detecting cells look different, but they work in a similar way. Sound makes the hairs on sound-detecting cells vibrate. Light makes molecules inside light-detecting cells change shape. Then both cells release chemicals, which make nerves send a signal to your brain.

Q

1. Draw a diagram to show a nerve cell leading to a muscle.
2. Suggest why separate brain cells are hard to see under the microscope.
3. Explain how cells can detect light and sound.
4. Make a flow chart explaining what happens to your sensory cells when you see the lights of a car coming towards you and you jump out of the way.

- Nerve cells carry electrical signals.
- Each long, thin cell connects with many other cells.
- They use chemicals to pass signals to other cells.
- Sensory cells send signals along nerves to your brain.

3.11 Tissues and organs

Objective
- Understand that cells work together to form tissues, organs, and whole organisms

Cells working together

Most of your body is made of muscle, nerves, fat, bone, and blood. Each of these is a **tissue**, which means a collection of similar cells. The cells in each tissue are specialised. They have different functions.

On average, males have more muscle than females, and less fat. You can change the amount of each tissue in your body by diet and exercise. Eating less makes fat cells smaller, and exercise can make muscle cells bigger.

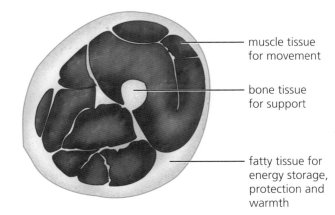

↑ The arrangement of bone, muscle, and fatty tissues in your arms and legs gives you strength and mobility.

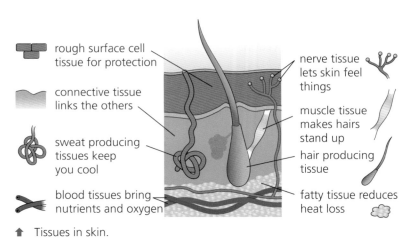

↑ Tissues in skin.

Teams of tissues

Tissues group together to form organs. Your skin is your biggest organ. Its different tissues help it to protect the rest of your body.

Tough surface cells keep micro-organisms out, and nerves let you sense things you should move away from – high or low temperatures, pressure, or pain. Hairs and fat keep you warm, and sweat-producing tissues cool you down. Connective tissues hold the rest together, and blood tissues bring nutrients and oxygen to keep all the other cells alive.

The tissues in your heart

As an organ like the heart develops, each type of cell moves into the right position and starts doing its job.

Scientists want to learn how this happens so they can mimic the process and grow complicated organs like hearts.

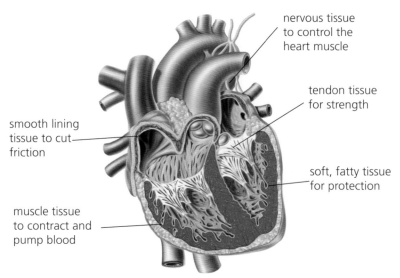

↑ Tissues in the heart.

Cells and organisms

Stem cells

Every cell is formed from existing cells that divide. When your body started to grow, your cells were all the same. They were all **stem cells**. As those stem cells divided, they produced every specialised cell in your body.

As your body developed, stem cells **differentiated**, which means they turned into the right specialised cells for each tissue. Their development was controlled by clues from their surroundings. These included chemical signals called **growth factors** made by other cells.

But how did these growth factors make the cells change?

Every cell in your body has a copy of the same set of genes. A muscle cell is different from a nerve cell because they have different genes turned on by the growth factors. Instructions from different genes make cells change their contents and their behaviour.

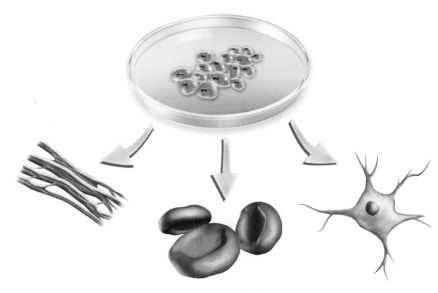

↑ Stem cells can produce every other type of cell.

Replacing your cells

Your red blood cells only last a few months. They can't divide to make new ones. New red blood cells are made by stem cells in your bone marrow.

Scientists are confident that they will learn to control stem cells. Then they will be able to use them to replace any damaged tissue.

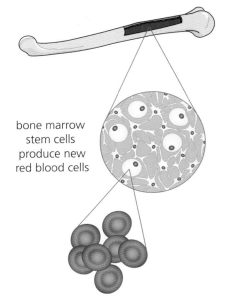

bone marrow stem cells produce new red blood cells

↑ Stem cells in bone marrow replace blood cells.

Q

1. Arrange these body parts in order of size: tissue, organ system, cell, and organ.
2. How can you tell whether something is a tissue or an organ?
3. Look at the tissues in skin. Suggest one type of tissue you would expect to find in every other organ.
4. Look at the tissues in the heart.
   a Name the two tissues that make the heart beat.
   b Name two other tissues inside the heart.
   c Name the tissue that protects the heart muscle.
5. Plants have different tissues too. Look at the root on page 45. How many different tissues can you see in it? What functions do the xylem and phloem tissues have?
6. Imagine you could see inside your arm. List all the tissues you would expect to find. Describe how their cells are specialised to suit their roles.

- Organs have tissues that suit their functions.
- Each tissue is made of a different type of cell.

Review 3.12

1. Scientists found these on the sea bed. They could be crystals, or they could be living things. How could they decide? [3]

2. These three images were taken using different microscopes.

Image	Size	Appearance
A	$\frac{1}{10}$ the width of a human cell	
B	$\frac{1}{1000}$ the width of a human cell	
C	variable – can be larger or smaller than a human cell	

   a Identify what each image shows. [3]
   b Which are living things? [1]

3. The same apple was photographed on the day it fell off a tree, and 2 weeks later.

   a Name the type of organism that caused its appearance to change. [1]
   b Explain why the breakdown of organic material is useful. [2]

4. Simisola used bacteria to turn some milk (pH 6.7) into yoghurt (pH 4.5).

   a Describe how bacteria convert milk into yoghurt. [2]
   b Simisola wants to compare the time taken to make yoghurt at different temperatures. List the variables she should control. [2]
   c How can she tell when the yoghurt is ready? [1]
   d Use her graph to estimate the temperature needed to produce yoghurt in the shortest time. [1]

   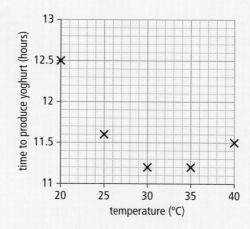

5. Ravi mixed some flour with a sugar solution to make dough. He added yeast to half the dough, then put both lots in measuring cylinders.

   He measured the starting height of the dough and the final height 2 hours later.

   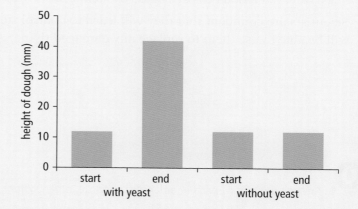

   a What do his results show? [2]
   b How does yeast cause this change? [2]
   c Suggest how he could increase the height the dough reached after 2 hours. [1]

6. Louis Pasteur invented pasteurisation.

   a Explain how food is pasteurised. [2]
   b How does pasteurisation affect the length of time milk stays fresh for? Give a reason for your answer. [2]

7 This apparatus was used to ferment fruit juice to make wine.

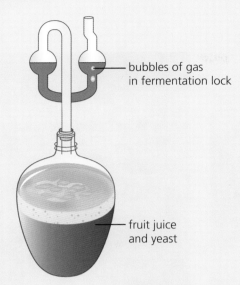

a The fermentation lock lets bubbles of gas escape. Explain why wine-makers can't just leave their bottles open. [2]
b Name the gas that escapes. [1]
c Why should the equipment be kept warm? [1]
d According to the graph, when did the amount of alcohol made stop rising? [1]

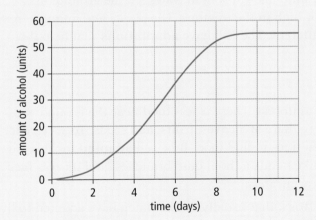

8 Identify each of the following cell structures and explain what they do.
a Without this structure, a plant cannot obtain the energy it needs. [2]
b This thin layer surrounds the cytoplasm. [2]
c Without these a plant could not stand upright. [2]
d This structure contains genes. [2]

9 The table shows the characteristics of three infectious diseases.

Disease	Micro-organism involved	Symptoms
A	virus in respiratory system	fevers, a runny nose, a sore throat, and sneezes
B	protozoa in blood	fevers, sweats, and muscle pains
C	bacteria in digestive system	fever, severe diarrhoea, headaches, and dehydration

a Suggest which diseases A, B, and C could be. [3]
b Suggest how each disease spreads to new victims. [3]

10 Inger investigated the connection between the number of root hairs on seedlings and the minerals in the soil they grew in.

Soil mineral content	Root hairs per mm	Average root hair length (mm)
high	48	160
low	60	560

a Describe Inger's results. [2]
b Explain how root hair cells help a plant to survive. [2]

11 Explain whether the following are types of cell, tissue, or organ.
a It is made of muscle, tendons, nerves, and fat, and pumps blood around the body. [2]
b It is a liquid full of red cells which circulates around your body. [2]
c Its nucleus and cytoplasm are squeezed to the side so that the space inside it can be used for storage. [2]
d It is made up of identical fibres which contract and pull on bones when they receive a signal from a nerve. [2]

4.1 Habitats

Different environments

Objective
- Describe how organisms are adapted to their habitat

↑ Rainforests are warm and wet.

↑ Deserts are hot and dry.

These two places are very different. Their **climates** are both hot and sunny but the rainforest is wetter. The plants here grow quickly all year round and there is plenty for animals to eat. Deserts get much less rain, so their plants grow more slowly and deserts do not produce so much food.

The place where a plant or animal lives is its **habitat**. Rainforests provide many different habitats for animals – in soil, in water, on the ground, in the darkness under trees, or high in the treetops. The plants and animals in each habitat can be very different. They all have **adaptations** that help them survive in their habitat.

Rainforest

This orang-utan lives in a rainforest. There is less food near the ground, so it stays in the treetops where there are leaves, fruits, and seeds. To survive it needs to cross gaps between trees to find food. It is well **adapted** for this – it has long arms, its feet act like hands to hold onto branches, and its eyes face forward to judge distances accurately.

↑ An orang-utan's adaptations help it move from tree to tree.

Forest air is damp so even frogs, that usually live in water, have adapted to live in the treetops. This flying frog eats insects. It uses suckers on its fingers and toes to climb, and webs between its toes to glide from tree to tree.

Desert

The water-holding frogs in Australia spend most of their lives underground. They grow fast when it rains, then fill their bladders with water and bury themselves in mud. They lie trapped underground for a year or more until it rains again. Then they pop out briefly to feed and breed. Native Australians used to dig them up and squeeze water from them to drink.

↑ Flying frog.

Food and water are scarce in deserts. Camels and oryx have adapted to life in the desert. They can walk long distances to find food and they can survive for months without drinking.

Most desert animals survive by hiding underground and saving water. They sleep during the day and come out at night to find food.

Antarctic

Antarctica is dark for 4 months of the year and mostly covered in snow. Few plants grow on land, but microscopic plants called plankton fill the oceans that surround the continent. Small shrimp-like animals called krill filter plankton from the water. There are more krill than any other type of animal on Earth. They feed most other animals in the Antarctic. First the krill are eaten by fish and some types of penguin, seal, and whale. Then sea birds and other penguins eat the fish.

In Antarctica there is less to eat in winter when plankton stop growing, but the animals here have adapted to these conditions. They can survive for months without food. Their bodies shrink as they use up their energy stores. Krill and fish make antifreeze to stop their bodies icing up. Penguins, seals, and whales have compact bodies and thick layers of fat to retain heat.

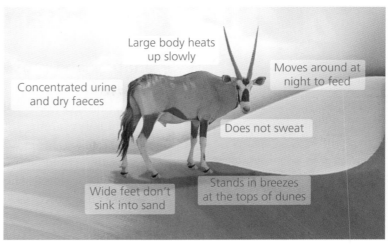

↑ These adaptations help oryx survive.

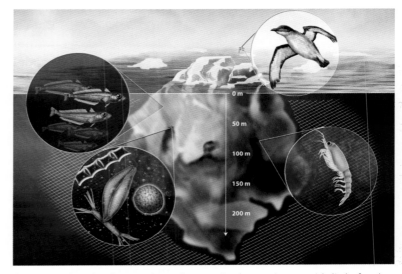

↑ Antarctic animals are adapted to survive long winters with little food.

1 Suggest a suitable habitat for each of these animals:

Kangaroo rats get all the water they need from the food they eat. They hide in burrows underground for most of the day.

Sloths eat leaves, which are not very nutritious. To save energy they hang upside down and sleep for most of the day. Microscopic plants live in their hair and turn it green.

Llamas have non-slip pads on the bottoms of their feet, and their thick woolly coats protect them from the wind. They also have larger lungs for their size than most animals.

2 Choose three different wild animals that live in your country. Describe each animal's habitat and the adaptations that help it survive there.

- Earth has regions with different climates.
- Within each region there are many habitats where different plants and animals live.
- Organisms have adaptations which help them to survive in their habitats.

4.2 Food chains

Predators and prey

Objective
- Draw and model simple food chains

Camarra is on holiday in Namibia. She sees herds of wild animals roaming freely. She'd like to get closer but knows that could be dangerous. The animals she can see are all **herbivores**, which means they only eat plants. Zebra and impala eat grass, and giraffes feed on acacia trees.

↑ African herbivores gathering to drink.

But **carnivores** are hiding in the undergrowth, just waiting for a chance to catch their next meal. These **predators** could attack at any moment.

The herbivores have sharp senses and they can run fast. They are hard to catch. Many carnivores hunt in packs to improve their success. These lions are feasting on their largest **prey** – a giraffe.

Once the lions have had their fill, vultures will pick the bones clean. They are **scavengers**. They only eat animals that are already dead.

↑ Three predators and their prey.

Producers and consumers

Herbivores get their nutrients by consuming plants so they are called **consumers**. Plants make all their own nutrients using energy from sunlight, so they are **producers**.

The amount of material in a plant (or any living creature) is called its **biomass**. The more plants there are, and the bigger they are, the more biomass there is for herbivores to feed on and the larger their populations can grow.

Every animal is part of at least one **food chain**. A food chain always begins with a producer. The arrows represent the flow of energy when herbivores consume plants and carnivores consume animals. Lions are the **top predators**, so they are the last link in this food chain.

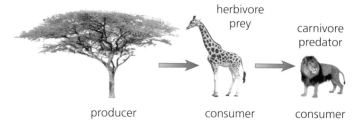

↑ This food chain shows how energy transfers from an acacia tree to a lion.

Acacia trees absorb energy from sunlight when they make their own nutrients. They use some of these nutrients to grow leaves - so leaves store some of the energy a plant absorbs. When giraffes eat the leaves, they use some of this energy to build muscles and fat. When the lions eat these, the energy passes to them. They store some of it in their own muscles and fat.

When the lions die, **decomposers** will break down their bodies and use this stored energy.

Living things in their environment

Food webs

Most animals eat more than one type of food so they are part of more than one food chain. The diagram shows how some of the lion's food chains link together to form a **food web**.

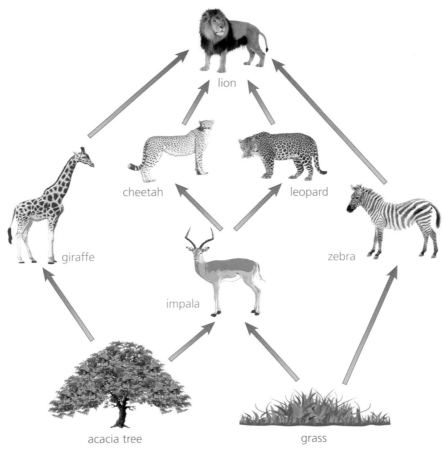

↑ The lion's food chains link together to form a food web.

Q

1. Write one food chain from the food web on this page. Label the herbivore and any carnivores in your food web.
2. Choose another food chain from the web. This time label any predators and their prey.
3. Pick another food chain. This time label the producer and consumers.
4. Explain what the arrows in your food chains represent.
5. Give an example to explain how a prey animal can pass energy to its predator.
6. Life is often described as 'a gift from the Sun'. Suggest why.
7. When less rain falls, the soil is drier and plants grow more slowly. Explain what effect this might have on the herbivores.
8. Imagine all of the carnivores in your food chains have been shot by poachers. Explain what would happen to the numbers of herbivores and plants.
9. Turn back to the section about the Antarctic on page 53. Use the information to draw one of the krill's food chains.

!

- A food chain shows how energy is passed from one organism to another.
- Food chains begin with producers which make their own food.
- Animals get their energy from producers so they are consumers.

4.3 Feeding ourselves

Objective
- Discuss the positive and negative effects humans have on food chains

Habitat destruction

This small orang-utan has a big problem. His forest home covered Borneo for thousands of years. Traditional farmers cultivated small patches for a year then moved on, so the forest grew back, but modern farming removes the trees for ever. People sell the wood and use the space to grow crops. More than half of the trees in Borneo have been cut down recently.

↑ This orange-utan's habitat is being destroyed.

This orang-utan shares the remaining forest with rhinos, leopards, elephants, native people, and uncountable numbers of smaller animals. When the plants at the beginning of each food chain are destroyed, there is nothing for animals to eat. If they can't move away, they die.

Pollution

To grow more food on less land, some farmers use chemicals. **Fertilisers** increase plant growth by supplying minerals. **Herbicides** stop weeds growing, and **insecticides** stop insects eating the plants.

If these chemicals are not used correctly, they cause **pollution**. Pollution damages plants and animals. When too much fertiliser runs into rivers, fish die. Insecticides kill useful insects like bees, and herbicides destroy wild plants. When any plant or animal disappears, more than one food chain may be disrupted.

Invasive plants and animals

In the 1970s, insects called large grain borers were accidently brought to Tanzania from Central America. Their favourite foods are maize and cassava. In their old home they had predators, but in Tanzania no animals ate them. Their population exploded. Now they have spread to other African countries and they destroy vast quantities of our food. **Invasive** plants and animals like these are damaging food chains all over the world.

← Large grain borers feed on maize and cassava.

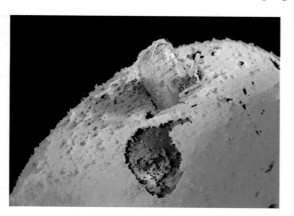

Living things in their environment

Planning for the future

The human population is still growing. There will be many more people to feed in the future. All our food relies on plant biomass. We can't use more biomass than plants can produce. Scientists estimate that humans already use more than a quarter of the total produced on land. Every other animal has to share what's left. If we could grow crop plants faster, there would be more to go round.

↑ If the cake was bigger, we could take more but leave the same for other animals.

Growing more producers

The producers in a food chain don't need to be plants. They can be micro-organisms. Tanks of algae can absorb much more sunlight than plants, so they grow much faster. We could use micro-organisms to make fuels and animal feeds, so we have more land left to grow food.

Half of the world's population lives in cities and the number is rising. If we built tall greenhouses in city centres, we could grow more food on less land. Plants grow faster indoors because they aren't damaged by bad weather or pests. We can also give them extra light and carbon dioxide to help them grow faster.

↑ High-rise greenhouses would let us grow more food in less space.

Avoiding pollution

Modern farms cover vast areas with one crop. They use huge amounts of fertiliser to replace the minerals crops take from the soil. Crops attract huge populations of insects. Farmers need to use insecticides to destroy them. They may be able to cut their use of chemicals by changing the way they farm.

Rice grows in flooded fields. If fish are added to the water, the rice and fish help each other. They both grow well without chemicals. The fish eat insects that would damage the rice and their waste products provide the minerals rice needs. In return, rice keeps the water clean and cool for the fish.

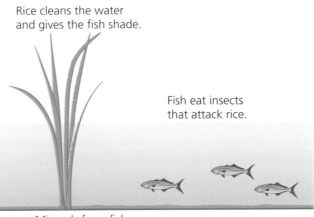

← Fish and rice help each other to grow well without chemicals.

Q

1 Describe three ways that food chains can be disrupted.
2 How can we get more food for ourselves without damaging other animals' food chains?
3 Explain some the benefits and drawbacks of the chemicals some farmers use.
4 Find one animal in your country that has been affected by human activity. Explain what humans did, and how it affected the animal.

- Humans destroy habitats when they clear land to grow crops.
- If farm chemicals aren't used properly they damage natural food chains.
- Invasive plants and animals can destroy existing food chains.
- We can reduce damage to other food chains by producing more food in less space and reducing our use of chemicals.

4.4 Changing the planet

Earth's atmosphere

Objective
- Discuss the positive and negative effects humans have on air pollution and ozone depletion

We couldn't survive without the atmosphere. The layer of air around the Earth allows us to breathe, lets plants make food, keeps Earth warm, and shields us from harmful radiation. It also moves water from the oceans to the land by making rain.

In the past the atmosphere seemed permanent and unchanging, but now we have evidence that that it is not. Human activities have damaged it.

↑ Earth's atmosphere protects living things.

Ozone

Ozone is one of the gases in the upper atmosphere. This ozone is vital to our survival. It stops harmful **ultraviolet** radiation reaching the Earth. Without this ozone, plant life would be damaged and millions of people could develop skin cancer.

Every year the layer of ozone over Antarctica gets very thin. The thin patch is called the ozone hole. What is causing this?

The parts of a fridge that get cold have a gas running through them. The first fridges used a harmful chemical. Nothing else would work. All that changed in the 1920s. A new gas was made. It seemed perfect for the job. It was the first **CFC**. Soon CFCs were being used for spray cans too.

50 years later scientists spotted the ozone hole, and realised CFCs were causing it.

Eventually governments reached international agreement. CFC use was stopped all over the world. Scientists hope the ozone layer will eventually recover.

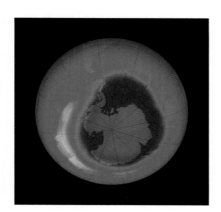

↑ There is least ozone in the dark blue section over Antarctica.

Acid rain

Fuels contain small amounts of **sulfur**. When this burns, an acidic gas escapes into the atmosphere. It dissolves in rainwater and makes **acid rain**. Acid rain damages trees and buildings, and harms life in lakes and rivers.

Once scientists spotted the problem they found ways of dealing with it. They now:

- take sulfur out of fuels before they are burned
- **neutralise** acidic gases before they can escape
- neutralise lakes and soil that are already damaged.

↑ These trees are being destroyed by acid rain.

Carbon dioxide

The amount of carbon dioxide in the atmosphere stayed about the same for most of human history. Plants take carbon dioxide out of the atmosphere, and all living things put it back when they respire.

When humans started burning coal, oil, and gas, things changed. These fuels add extra carbon dioxide to the atmosphere when they burn – more than plants can use up. Cutting down forests makes the problem even worse.

As carbon dioxide levels rise, so does Earth's average temperature. Carbon dioxide works like the glass in a **greenhouse**. It lets light from the Sun through, but stops some of Earth's heat escaping. The more carbon dioxide there is in the atmosphere, the hotter Earth will get. This change in Earth's temperature is called **global warming**.

Light from the Sun passes through and heats the ground.

Some of the heat from the ground cannot escape.

← Carbon dioxide acts like the glass in a greenhouse.

International agreement to cut carbon dioxide release has not been very successful. Scientists are planning new ways to take carbon dioxide out of the atmosphere. If that doesn't work they could use mirrors to reflect heat away, or add chemicals to the atmosphere that make clouds to keep out the Sun.

← The Earth could be cooled by using mirrors to reflect heat or by adding chemicals to the atmosphere that make clouds form.

Q
1. Explain why humans should worry about changes in the atmosphere.
2. Describe one negative and one positive effect that human activities have had on the ozone layer.
3. Describe one negative and one positive effect that humans have had on acid rain.
4. Scientists think we should try to stop carbon dioxide levels rising. Why do they think this?
5. Describe two ways that we can stop the amount of carbon dioxide in the atmosphere rising.
6. How could clouds and mirrors be used to stop Earth getting warmer?

- A layer of ozone in the atmosphere blocks ultraviolet radiation from reaching Earth.
- CFCs used in the past have destroyed ozone, and put plant and animal life at risk.
- Carbon dioxide keeps Earth warm and allows plants to grow.
- Burning fossil fuels and cutting down forests increase the carbon dioxide in the atmosphere and cause global warming.
- Burning fuels also produce acid rain, which damages plants, aquatic life, and buildings.

4.5 Preventing extinction

Objective
- Understand how we can conserve endangered plants and animals

Lost animals

In December 2006, scientists spent 6 weeks searching China's Yangtze River for dolphins. They didn't find any. The type of dolphin that used to live there had all died out – they had become **extinct**. Most dolphins live in the sea, but this was a river dolphin. It was a unique type of dolphin found nowhere else on Earth.

Now alligators, turtles, and porpoises are also struggling to survive in the Yangtze, and fish numbers have dropped dramatically. Scientists blame pollution. Many think the river will be impossible to clean up.

↑ This river dolphin became extinct.

➡ Criminals shoot rhinos so they can saw off their tusks and sell them.

Providing sanctuaries

Saving animals from extinction is called **conservation** and is carried out mainly by charities. One animal that charities want to save is the Indian rhino.

Rhinos are endangered because criminals shoot them and cut off their horns to sell. The horns are used to make traditional Asian medicines. However, there's no evidence that the medicines do any good.

Conservationists have moved some rhinos to **wildlife sanctuaries**. These are special areas where the animals can be protected. Charities pay people to guard them, and pay for the equipment they need to do the job.

Saving habitats

It takes a lot of land to support large predators like Bengal tigers. They need to roam vast areas of forest to find enough prey. When land is cleared to build roads or houses their territory is cut up. The pieces of land left behind can be too small for tigers.

To save the tigers, conservationists need to save their habitat. This has an added advantage. It saves all the other plants and animals that live there too.

Conservationists work with local people. They find them other ways of making money so they don't need to cut down the trees.

Removing predators

When early sailors visited new islands, they accidently took rats there that were living on their ships. The birds that nested on these islands were defenceless. Rats ate most of their eggs, and many types of bird became extinct. The islands around New Zealand were badly affected. Some had birds that were found nowhere else in the world.

Since the 1960s people have attempted to restore life on many islands. The first step is to spread rat poison all over them. Once the rats are dead the birds can reproduce safely. Birds that are close to extinction on the mainland can also be taken to the rat-free islands to recover.

⬆ Conservationists hope to save Bengal tigers.

Captive breeding

The Arabian oryx was hunted to extinction in the wild. There were just a few left in zoos. When there are only a few animals left, a **captive breeding program** is the only way to save them.

One American zoo had nine oryx. They started to breed them. When they had 200, they sent some to other zoos to do the same. Eventually herds of the animals were sent back to their original homes. Most of them have survived.

⬆ Oryx were saved by breeding them in captivity.

Q
1. Describe three things that can make an animal extinct.
2. How can wildlife sanctuaries help save animals from extinction?
3. Describe a habitat in your country that is being damaged by human activities. Suggest what could be done to save some of the plants and animals that live there.
4. Explain why conservationists breed endangered species in captivity.

- We can save plants and animals from extinction by reducing pollution, setting up wildlife sanctuaries, saving habitats, removing predators, and using captive breeding programs.

4.6 Obtaining energy

Objectives
- Recognise a variety of sources of energy
- Distinguish between renewable and non-renewable energy sources

Fossil fuels

We need energy for everything we do. Our own energy comes from respiration fuelled by food. We use other energy sources to generate electricity and run cars, trucks, trains, and planes.

Most power stations burn **fossil fuels** – coal, oil, or gas – to make electricity. These fossil fuels are **non-renewable**. They took millions of years to form from dead plants and animals. They cannot be replaced and will run out fairly soon. Oil is particularly valuable. It is used to make petrol, diesel, paraffin, and fuel oil for transport.

Most power stations use the energy from fossil fuels to make electricity.

Burning fossil fuels adds to global warming by putting carbon dioxide into the atmosphere, so it may be a good thing that we won't be able to use them for long.

Solar energy

Solar cells provide electricity for this mobile phone mast.

Decomposing animal manure gives off methane gas, which is a useful fuel for cooking.

Some energy sources will never run out. They are constantly replaced by the Sun, so they are **renewable**. Energy from the Sun is called **solar energy**.

Solar cells turn the Sun's energy into electricity. They are used in calculators, on illuminated road signs, and on buildings. They can only produce small amounts of electricity, but they are very useful in isolated areas. Energy from solar cells can be stored in batteries to use when it is too dark or cloudy for the solar cells to work.

Biofuels

Biofuels are made from plants. Plants produce a new crop every year, so they are renewable. Biofuels can be replaced as fast as they are used.

Plants use solar energy to grow and some of that energy is stored in their tissues. We can use plant materials to make liquid fuels for cars. We can also burn them in power stations to generate electricity.

Cows eat a lot of grass, but they don't extract all its energy. There is still some energy left in their dung. When dung is left to decompose, it makes a gas that can be used as a fuel. A similar gas is made when waste decomposes in rubbish dumps.

Wind and water

Solar energy evaporates water which then falls as rain. When rain falls on high places, it runs downhill. The Sun's heat also makes winds blow. Anything moving has energy, so the moving water and wind can generate electricity.

These energy sources can be expensive to begin with because turbines and dams cost a lot to build.

Geothermal energy

↑ These turbines convert the wind's energy to electricity.

↑ Dams control the flow of water to generate electricity.

← The energy in this steam comes from hot rocks underground.

Geothermal energy is energy from hot rocks under the Earth's surface. The heat energy converts water to steam, which can be used to make electricity. Geothermal energy can only be used in places like East Africa where there are hot rocks close to the Earth's surface.

This is a very useful energy source. The equipment needed to collect the energy doesn't cost much and the energy is available all the time.

Q
1 Explain the difference between renewable and non-renewable sources.
2 Give three examples of each type of energy source.
3 Which renewable energy source does not depend on the Sun?
4 Suggest which renewable source might be best in each of these places:
  a an isolated village in Africa
  b a house in the city
  c a volcanic area with hot rocks near the surface
  d a town in a mountainous region with fast-flowing rivers
  e a village in a hilly area where there are no rivers.

- Fossil fuels are non-renewable energy sources. They cannot be replaced and will eventually run out.
- Renewable sources will not run out because they are constantly replaced. They include sunlight, biofuels, wind, water, and geothermal energy.

Extension 4.7

Growing fuels

Objectives
- Recognise some sources of renewable fuels
- Recognise that each biofuel has advantages and disadvantages

Reducing pollution

Burning fossil fuels adds carbon dioxide to the atmosphere. Even if fossil fuels weren't running out, we should try to stop burning them.

Biofuels also release carbon dioxide when they burn, but the next crop takes the carbon dioxide back out of the air again as it grows. The diagram shows how these crops recycle carbon dioxide. Fuels that recycle carbon dioxide are called **carbon-neutral** fuels because they don't change the amount of carbon dioxide in the air.

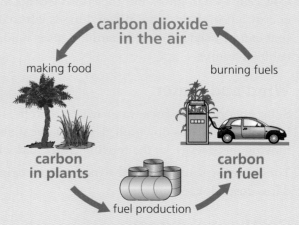

↑ Biofuels should be carbon-neutral.

There are two main biofuels, **biodiesel** and **bioethanol**. Biodiesel is made from plant oils and bioethanol is made from sugars. The 'bio' means they are made from plants instead of fossil fuels.

↓ The palm oil in these fruits can be used in food or biofuels.

Palm oil

Oil palms grow in Malaysia where there used to be rainforest. Traditional farmers used to grow food for their families and local markets there, but now modern farmers sell the palm oil all over the world. The tropical conditions here are perfect for growing palm oil.

This oil is used in 1 in 10 supermarket items, including crisps and bread. Now it is also being used to make biodiesel.

Palm oil production makes huge profits for many large companies, but scientists have evidence that the process isn't really carbon neutral.

Hidden costs

This fire was started deliberately. Farmers will grow oil palms on the empty land. Clearing forest by burning it like this releases huge amounts of carbon dioxide. Growing the oil palms will not take in all this carbon dioxide. The extra carbon dioxide will stay in the atmosphere.

Another problem is that energy is needed to produce biofuels. It takes more energy than is needed to make petrol from oil. Most of the energy for making biofuels comes from burning fossil fuels. So switching to biofuels doesn't always reduce pollution.

↑ Deforestation releases masses of carbon dioxide into the air.

Living things in their environment

Bioethanol

Up to 20% of a sugar cane plant is sugar. Yeasts can use this sugar to make alcohol. The type of alcohol they make is called ethanol. When this ethanol is produced as a fuel it is called bioethanol. The 'bio' shows that it is renewable because it is made by living things.

If all a plant's biomass could be turned into sugar, it could all be turned into bioethanol by yeast. But plant cell walls are tough. Only a few decomposers can break them down.

One decomposer that is very effective is this fungus. You can't see most of it. Its hyphae are inside the tree trunk. Their enzymes turn plant cell walls into a sugary mush. We can use these enzymes to turn fast-growing plants into sugars that yeasts can use. So we can use these plants to make ethanol, and save the sugar in sugar cane for humans to eat.

↑ This fungus is using enzymes to turn the inside of the tree to mush.

Bioethanol production is cheaper in warm countries where sunlight can be used to separate the fuel produced from water. The fungal enzymes are expensive, but they are expected to get cheaper in the future.

Algal oil

Algae grow fast, and they can contain lots of oil – up 60% of their biomass. They could grow in contaminated water, and use up the carbon dioxide power stations make. So they could clean up pollution and make fuel at the same time.

Algae could be grown out in the open, in huge ponds. But scientists don't think this would work well. The ponds would soon be infected by microbes from the air. Instead scientists want to grow algae in **photobioreactors**. These are tubes surrounded by artificial light. They make it easy to give the algae the right amounts of light, carbon dioxide, and minerals.

Photobioreactors are expensive, but could be built anywhere. They would not use up valuable farm land like the crops grown to make bioethanol.

↑ Algae grow fast in photobioreactors.

1 Name two different biofuels. Where do they come from?
2 Suggest two reasons why we should use biofuels instead of petrol.
3 Explain why biofuels should be carbon neutral.
4 Give a reason why biodiesel from palm oil is not carbon neutral.
5 Explain how fungi help us to make bioethanol.
6 Give one advantage and one disadvantage of each of these types of fuel:
   - bioethanol
   - biodiesel

- To help the environment, biofuels should release less carbon dioxide than petrol or diesel.
- Making biofuels can do unintended damage to the environment.
- Newer biofuels should do less damage to the environment.

Review 4.8

1 The arctic fox lives in cold climates and the fennec fox lives in the desert.

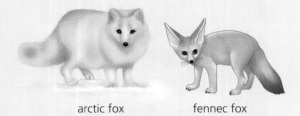

arctic fox   fennec fox

   a Describe two differences between the foxes. [2]
   b Suggest which fox has the thickest layer of fat under its skin. [1]
   c Give two reasons why the fennec fox would not survive in the Arctic. [1]

2 Karim is studying food chains.

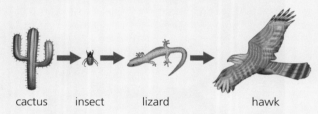

cactus   insect   lizard   hawk

   a Choose one organism from this food chain that fits each of the following categories: carnivore, herbivore, producer, consumer, predator, and prey. [6]
   b Foxes eat lizards but not hawks. Explain how a rise in the fox population could affect the hawk's ability to raise its offspring. [2]
   c What would happen to the cacti if all the lizards were killed by a disease? Explain why. [2]

3 Cactus plants are adapted to live in deserts.

Describe two adaptations that help these plants survive months without rain. [2]

4 The graph shows estimates of the size of an orange-utan population in different years.

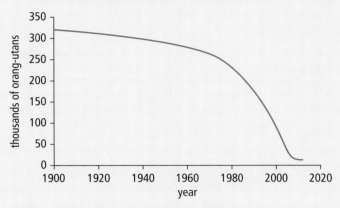

   a Describe how the population has changed since 1900. [2]
   b Suggest a reason for the sudden change between 1960 and 1980. [1]
   c Suggest a reason why numbers have fallen more slowly in recent years. [2]

5 Wheat is one of the world's main crops. The Russian wheat aphid feeds on wheat and cuts the amount farmers can harvest. Since this pest was spotted in Russia in 1912, it has spread to many different countries.

The graph shows how their numbers changed on one farm.

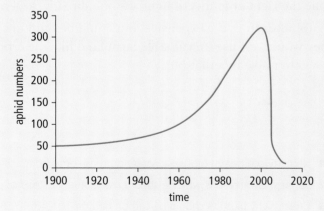

   a Sketch the shape of the graph. [1]
   b Ladybirds eat aphids. Add a second line on your graph to suggest how the number of ladybirds on the farm may have changed. [2]
   c Suggest a reason for the sudden fall in the number of aphids. [1]

Living things in their environment

6 Timmy tested the effect acid rain has on plants. He planted identical seeds into two separate pots and gave both pots the same volume of water every day. One pot received pure water and the other received acidic water. Both pots were left near the same window.

Water used	Appearance	Height (cm)
acidic	a few brown leaves	4.0
pure	lots of green leaves	9.5

a List three variables that Timmy controlled. [3]

b What can he conclude about the effect acid rain has on plant growth? [2]

7 Carbon dioxide concentrations in the atmosphere are measured in parts per million (ppm). The graph shows how they changed between 1960 and 2000.

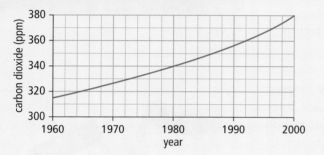

a Describe in detail how the concentration of carbon dioxide in the atmosphere changed. [3]

b Suggest two things that could have produced this change. [2]

8 Polar bears eat mainly seals. They catch them by waiting at holes in the frozen sea where seals come up to breathe.

Every summer the ice melts. Seals are then too hard to catch in the open sea, so polar bears live off their fat until the sea freezes over again.

Zosia wants to find out how climate change will affect polar bears.

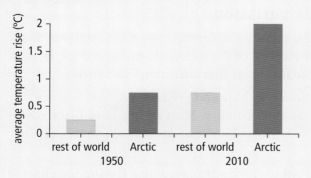

a What do Zosia's data show? [2]

b How will this affect the time it takes for the ice to melt each year? [1]

c Suggest how this could affect the number of seals polar bears catch. [1]

9 The pie chart shows the types of energy used to generate electricity in one part of the USA.

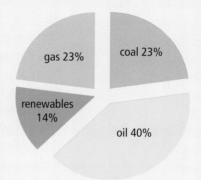

a Calculate the total percentage of energy obtained from fossil fuels. [1]

b Explain how and why this figure will change in the future. [2]

c List four sources of renewable energy. [4]

d Explain why switching to renewable energy sources could reduce global warming. [2]

10 A scientist compared the amount of ultraviolet radiation reaching Antarctica one spring with the ozone levels in the atmosphere.

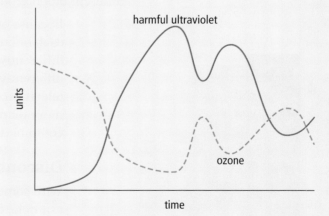

a Compare the two graphs. [2]

b Explain the difference between the two graphs. [1]

c Give one reason why scientists want to limit the amount of ultraviolet radiation reaching the Earth. [1]

5.1 Variation

Objective
- Investigate variation

Differences

Most of us can recognise hundreds of people – friends, family, people we only see occasionally, and celebrities on TV. The shapes of their faces differ and their hair, eye, and skin colours vary.

You can spot people you know from a long way off. They have different heights and builds, individual voices, and distinct ways of walking. We have thousands of features like this which vary. The differences between us are called **variation**.

⬆ Every individual is unique.

Hidden variation

Some differences are easy to spot, but most variation is hidden. For example, you might be able to hold your breath for longer than most people, or maybe your heart beats more times per minute. Nobody would know this from looking at you. Your blood looks the same as your friend's, but it may belong to a different blood group.

Because of variation, each of our bodies works differently. We also behave differently – we prefer different television programmes and laugh at different jokes. There are more than 7 billion people in the world. We all have different combinations of features, so we are all unique.

⬆ Everyone has one of the 4 main blood types.

Discontinuous variation

With some features, there are only a few possible variations. For example, everyone has one of the four main blood groups: O, A, B, or AB. Features like this show **discontinuous variation**.

There's lots of variation in hair and eye colours, but they can be sorted into a few main colours. Most people have black hair and brown eyes but other colours are common in some parts of the world.

Discontinuous variation can be displayed on a **frequency chart**. The height of each bar shows the number of people in each **category**.

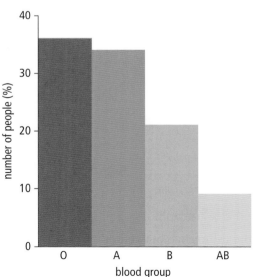

⬆ This frequency chart shows the distribution of blood types in Egypt.

Continuous variation

Body measurements, like height or body mass, show **continuous variation**. They can take any value within a certain **range**.

To make a frequency chart, you collect the values together in equal-sized groups. Then you plot the number in each group.

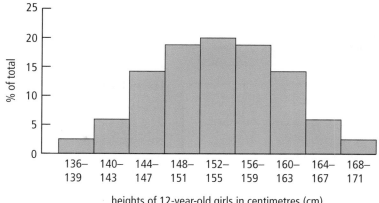

Frequency charts for continuous variables often have this sort of shape.

Unique differences

Everybody's fingerprint has a unique pattern of lines. Even identical twins have slightly different fingerprints. Every person also has:

- a different shaped face
- a different pattern on the coloured part of each eye (iris)
- a unique voiceprint
- a distinctive way of walking.

Any of these features could be used to check their identity.

Biometrics

Using features like fingerprints to identify people is called **biometrics**. Many countries use biometric data on passports.

Security systems like those used at airports need to be fast, reliable, and as cheap as possible. It can be difficult to decide which biometric features to use. Each has advantages and disadvantages.

Iris patterns vary more than fingerprints and face shapes. They provide tighter security, but people have to stop to look into an iris scanner. Face shapes can be detected from further away. They could be used to make a door open as soon as someone walked up to it. Fingerprint scanners are the cheapest, but they don't work well with dirty or damaged fingers.

This scanner uses fingerprint patterns to check people's identity.

Q

1. Look at the people in the photograph on page 68. List five visible features that vary from person to person.
2. Suppose you could carry out some tests on each person. Describe five features you would expect to show variation.
3. Variation can be continuous or discontinuous. Explain what is different about these two types of variation.
4. Choose three visible human features that show continuous variation.
5. Explain what a frequency chart can tell you about a discontinuous variable like blood group.
6. Explain why detectives often look for fingerprints at the scene of a crime.
7. **Extension:** choose the best sort of biometric data that could be used to control access to a computer. Remember to explain your decision.

- Everyone has a different combination of features.
- Differences between individuals are called variation.
- Blood groups show discontinuous variation.
- Height and body mass show continuous variation.
- Everyone has some features that are unique.

Extension 5.2

Causes of variation

Objective
- Understand why individuals are different

Inherited features

Rachel has two older sisters and her mum is about to give birth again. The three girls look different, but they have all inherited some of each parent's features.

Some features are present when we are born. For example, Rachel's eye colour and blood group won't change as she gets older. Other features develop as we grow and age. Rachel is still growing – she can't be sure how tall she will get, or what her final body mass will be.

⬆ Each girl inherited some features from each parent.

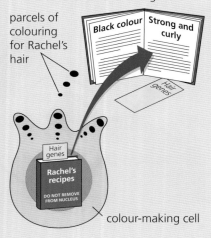

⬇ Each type of cell uses the instructions in different genes.

Genes

Your features depend on the **genes** you inherit from your parents. There is a copy of these genes in the nucleus of every cell in your body. They act as your body's recipe book. Genes tell your cells what to do. Your cells form tissues and organs, so instructions in your genes shape your whole body.

Most cells only have certain functions. They don't use every gene. Some of Rachel's cells just produce colour for her hair. They follow the instructions in her hair colour genes and make her hair black.

Some genes are the same in everyone, so they produce the same body parts in everyone. However, there are many different genes that control hair colour. Rachel's hair would be a different colour if she had inherited different genes. Variation caused by genes is **inherited variation**.

The environment

This miniature tree is called a bonsai. It has the same genes as a full-sized tree, but the shallow tray stops its roots growing.

The tree's small size is not caused by genes. Plants that don't get enough light, water, and minerals don't grow as big as those that do. This is an example of **environmental variation**.

⬆ These trees have the same genes but very different environments.

Variation and classification

Identical twins

Humans and other animals are affected by their environments too. Your environment doesn't just mean your surroundings. Everything you eat, drink, do, and learn affects you in some way.

Samuel and Moses are identical twins. They inherited the same genes from their parents. The twins were hard to tell apart when they were babies. Samuel was ill last year. He lost a lot of weight, but he still plays football. Moses prefers boxing. He has built up his muscles by lifting weights. You can see differences between the twins – this is environmental variation.

Most of your features are affected by both your genes and your environment. You can inherit the genes for a fit, athletic body; but you won't have one unless you eat well and train hard.

Identical twins have the same genes but environmental variation can make them different.

Changing environments

Between 1985 and 2005, the average height of Chinese teenagers increased by 6 cm. This isn't because their genes have changed. Young people have a better standard of living now and more nutritious food, so more of them reach the maximum height their genes can produce.

Your genes and environment interact throughout your life to control the way you grow and develop.

On average, teenagers' heights increase when they eat more nutritious food.

Q

1. List two visible features that Rachel has inherited from her mother.
2. Explain how a gene can influence what you look like.
3. List two of Rachel's features that won't change after she is born because they are not affected by her environment.
4. People whose diets lack certain minerals are often shorter than average. Is this an example of genetic or environmental variation?
5. The lower leaves on a tree are often thinner and paler than those near the top. What difference in their environments could cause this?
6. If a seed from a bonsai was planted outside in a field, would it grow into another miniature tree? Explain your answer.
7. Choose three features from this list that are controlled by both genes and the environment: language/s spoken, fingerprint pattern, blood group, skill at football, eye colour, body mass, and health.

- Everyone inherits a unique set of genes from their parents.
- Genes shape your body by controlling what your cells do.
- Your environment also influences most of your features.

5.3 Species

Cats and dogs

Objective
- Understand what is meant by a species

➜ Cats and dogs both show a lot of variation but each species shares some important characteristics.

These animals look very different, but you can tell which animals are cats and which are dogs. The cats and dogs belong to separate **species**. The cats' faces have similar shapes, while all the dogs' faces are different.

A **species** is a group of organisms that share the same characteristics. Nearly 2 million species have been recorded and biologists estimate that at least 5 million species exist.

Distinguishing species

How can we decide which animals belong to the same species? Sometimes members of the same species can look very different. On the other hand, members of different species can look the same.

Scientists use this definition to clarify whether living things are the same species:

- Members of a species can breed with each other and produce offspring that are able to breed.

↑ A liger is a hybrid.

Members of different species can't usually breed with each other, but there are exceptions. A liger is the offspring of a male lion and a female tiger. Its parents are from different species, so the liger is a **hybrid**. Hybrids are usually **infertile**. They can't produce offspring of their own.

Unfortunately breeding experiments take a long time and animals don't always cooperate, so decisions about species can be difficult.

Naming species

Clouded leopards hunt alone – mainly at night – and spend most of their time in the treetops. They are rarely seen close up. They roam Southern China as well as islands like Borneo. Their Chinese name means 'mint leopard' but the Malaysians on Borneo call them 'tree tigers'. It's confusing when the same animal has different names.

↑ Clouded leopards are small but they share some characteristics with lions and tigers.

To avoid this confusion, scientists give each species a unique two-part **Latin** name. Similar species are given the same first name. For example, lions and tigers are both large and they both roar. They share the first name *Panthera*. Pet cats purr, along with some small wild cats. They share the first name *Felis*.

Variation and classification

The second name identifies each individual species. So lions are *Panthera leo* and tigers are *Panthera tigris*.

It was difficult to decide which first name clouded leopards should have. They purr and roar. Their skulls look like a big cat's but are much smaller. Scientists solved the problem by giving them a first name of their own – *Neofelis* – meaning new small cat. Their second name is *nebulosa*, which is Latin for cloudy.

A new species

For a long time clouded leopards on Borneo and mainland China shared the same name – *Neofelis nebulosa*. Then scientists checked to see if they really were the same species.

Instead of trying to breed the clouded leopards, scientists compared the genes in their cells. They found that they were as different as the genes of lions and tigers. The animals on Borneo became a separate species and were given a new name – *Neofelis diardi*.

Genetic evidence lets scientists decide whether animals are the same species or just look similar.

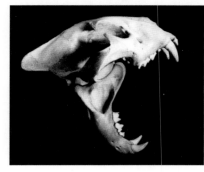

↑ Clouded leopards have similar skulls to lions and tigers.

↑ Clouded leopards in China and Borneo look the same, but their genes show that they are different species.

Q

1. What is a species?
2. Do members of a species always look similar?
3. Mules are the offspring of female horses and male donkeys. They are stronger and more intelligent than horses. Explain why they aren't a separate species.
4. Wild cats can look very different. How could you prove that a male and female belong to the same species?
5. Why can't we be sure how many species there are in the world?
6. Suppose you discovered a large white cat that roars. How would you name it scientifically?

!

- Members of the same species have similar characteristics.
- They can breed with each other and produce fertile offspring.
- Each species has a unique two-part scientific name.

5.4 Classification

Objectives
- Understand what classification involves
- Classify invertebrates into their major groups

Putting animals in groups

How do you store data about 2 million species, and how can you make sure they all have different names? To make these jobs easier scientists put similar species together. They use their similarities and differences to sort them into groups and sub-groups. This is **classification**. Scientists everywhere use the same system.

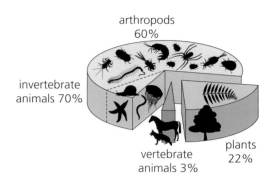

↑ Most species are animals, and most animals are a type of invertebrate called an arthropod.

Most species can be classified as plants or animals. The difference between them is that animals feed on other living things while plants make their own food.

Vertebrates and invertebrates

Plants and animals are divided into smaller subgroups. For animals, the first division is between **vertebrates** and **invertebrates**. Vertebrates have bony skeletons with backbones. Invertebrates don't.

Sometimes it's hard to tell which is which. A roundworm and snake look similar. The roundworm is an invertebrate, but the snake has a backbone – it's a vertebrate.

↑ These animals belong to completely different groups. The roundworm is an invertebrate but the snake has a backbone so it's a vertebrate.

Groups of invertebrates

Vertebrates and invertebrates can both be sorted into smaller subgroups with different characteristics.

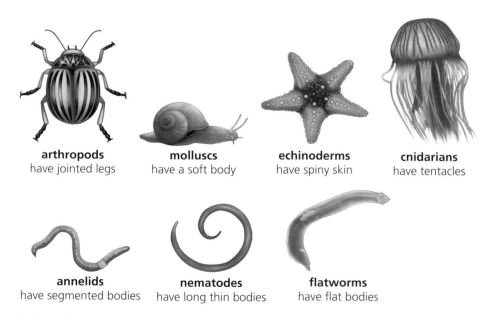

arthropods have jointed legs
molluscs have a soft body
echinoderms have spiny skin
cnidarians have tentacles
annelids have segmented bodies
nematodes have long thin bodies
flatworms have flat bodies

↑ Invertebrate groups.

Arthropods

Most invertebrate species belong to one subgroup – the **arthropods**. Their group name means 'animals with jointed legs'. There are more than a million arthropod species, so they need to be split into smaller subgroups.

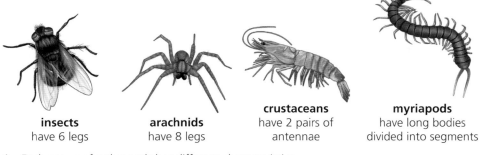

| **insects** | **arachnids** | **crustaceans** | **myriapods** |
| have 6 legs | have 8 legs | have 2 pairs of antennae | have long bodies divided into segments |

⬆ Each group of arthropods has different characteristics.

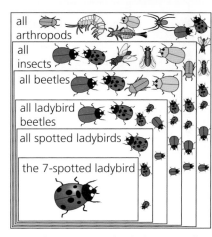

⬆ Arthropods are divided into smaller and smaller groups until each group contains a single species.

Insects

Most arthropod species are **insects** – animals with six legs. Nearly a million insect species have been named. A third of all insect species have tough covers over their wings – they are **beetles**.

Five thousand of the beetle species are called **ladybirds**. They have brightly coloured, rounded bodies. Many ladybirds are red with black spots, but only one species has exactly seven spots – *Coccinella septempunctata*.

Using a key

Having all these subgroups makes it easy to identify each species using a key. A key uses simple questions to determine which groups and subgroups a species belongs to. *Coccinella septempunctata* belongs to each of these groups: arthropods, insects, beetles, ladybirds and spotted ladybirds.

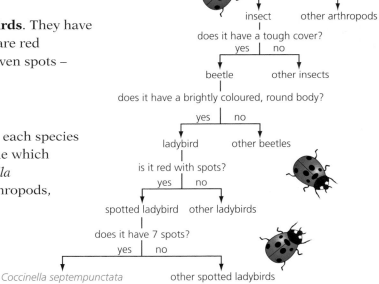

⬆ To use a key, start at the top and follow the arrows as you answer each question.

Q

1. Use the pie chart to find the percentage of species classified as animals.
2. Which group has more species – vertebrates or invertebrates?
3. How are vertebrates different from invertebrates?
4. Which group of invertebrates has most species?
5. Identify what type of invertebrate each of these animals is. Explain how you know.

!

- Species are classified by sorting them into groups.
- All animals are either vertebrates with backbones, or invertebrates.
- Invertebrates are classified into many groups and subgroups.

5.5 Vertebrates

Making groups

There are fewer vertebrate species than invertebrate species, but it is still helpful to classify them. Vertebrate species form five main groups. The features each group shares are shown in the chart below.

Objective
- Classify vertebrates into their major groups

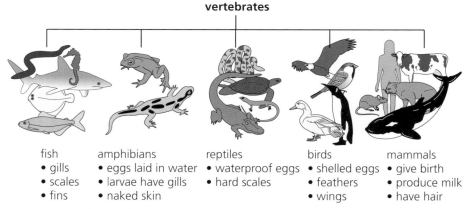

fish
- gills
- scales
- fins

amphibians
- eggs laid in water
- larvae have gills
- naked skin

reptiles
- waterproof eggs
- hard scales

birds
- shelled eggs
- feathers
- wings

mammals
- give birth
- produce milk
- have hair

↑ Vertebrates have different body coverings, different ways of reproducing, and different ways of taking in oxygen.

Fish are only found in water. Their gills don't work in air.

Amphibians can breathe air but they only reproduce in water. Their young have gills.

Reptiles, birds, and mammals can live in drier places. They have waterproof body coverings and they can reproduce without water.

Mammals give birth to live young and feed them on milk, but other groups lay eggs.

Warm blood

Mammals and birds keep their body temperature around 37 °C – they are warm blooded. It takes a lot of energy to keep warm, so they need to eat more for their size than species in the other three vertebrate groups. Fish, amphibians, and reptiles are cold blooded. Their temperatures usually stay the same as their surroundings, but they can warm up by basking in the Sun.

Amphibians and reptiles

Amphibians and reptiles can be hard to distinguish from a distance. Both these animals have four legs and a tail. The salamander is an amphibian. Its skin is smooth and moist and it lives in cool, damp places. Its young develop in water so they have gills instead of lungs.

The gecko is a reptile. Tough scales protect its skin and it lays its eggs on land. Some reptiles live in or near water, but geckos like hotter, drier environments.

↑ Unlike most mammals, bats let their body temperature drop down at night to match their surroundings. This saves energy so they don't have to eat so much.

↑ A salamander and gecko look similar.

Variation and classification

Unusual mammals

Whales are unusual mammals. They have very little hair and their bodies are streamlined, but like other mammals they give birth to live young. Mothers push their newborn calves to the surface to take their first breath. Their offspring feed on milk for at least 6 months.

← Like all mammals, whales feed their young on milk.

Difficult to classify

Some species have characteristics from more than one vertebrate group. Their body covering is usually the best guide to the group they belong in.

Olm
- has gills
- covered in smoooth skin
- lives entirely in water

Platypus
- has a double layer of fur
- lays eggs
- offspring fed on milk

Caecilian
- smooth skinned
- lives in underground burrows
- gives birth to live young

Armadillo
- underside covered with soft fur
- low body temperature
- feeds young on milk

↑ Some vertebrates don't share all the characteristics of their group.

1 Use the evidence below to classify each animal:
   a Dolphins feed their young on milk.
   b Snakes have dry scaly skin.
   c Sea turtles lay their eggs in the sand.
   d Sea horses breathe using gills.
   e Penguins lay shelled eggs.
   f Whales lose their hair before they are born.
   g Komodo dragons have dry scaly skin.

2 Bats are flying mammals. They sleep during the day, so they are rarely seen. List the features you would expect them to have.

3 Think of two very different vertebrates that live in water. Which vertebrate groups do they belong to?

4 Think of two very different vertebrates that live on land. Which vertebrate groups do they belong to?

5 A pet snake only needs feeding once a week. Most cats and dogs eat twice a day. Explain why cats and dogs need more food.

6 Explain what makes each of the following vertebrate animals different from most of their group: whale, bat, snake, and penguin.

7 Classify the following animals and describe a feature they have that doesn't fit their group: olm, platypus, caecilian, and armadillo.

- Vertebrates are classified as fish, amphibians, reptiles, birds, or mammals.

5.6 Classification of plants

Objective
- Classify plants into their major groups

Flowering plants

Most plants are **flowering plants** like those on page 8. Flowers come in a huge variety of shapes and colours. Closely related plants have very similar flowers. This makes it easy to classify flowering plants into groups and subgroups.

Grasses and trees are flowering plants too, but their flowers are usually small and dull.

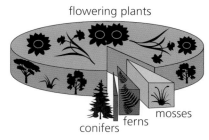

↑ Most plant species are flowering plants.

Plants without flowers

The three plants in the photograph are different. They are not flowering plants.

The moss on the ground has no roots to take in water. The other plants have roots to take in water and veins in their stems to carry the water up to their leaves. They can grow taller than the moss.

Mosses and ferns grow in damp, shady places. They reproduce by making **spores** and can only do this successfully when they are wet. Spores are smaller than most seeds and don't last as long. Moss spores form at the ends of small stalks, while fern spores form on the undersides of their leaves.

The trees are **conifers**. Conifers and flowering plants can reproduce in dry conditions. They both make seeds but conifers have cones instead of flowers. Seeds can survive for thousands of years if they are cold and dry.

↑ Moss, ferns and conifers reproduce without flowers.

↑ Conifer seeds are produced in cones.

↑ The red capsules on the moss contain spores.

↑ Fern spores form on the undersides of leaves.

Life in the oceans

In the oceans plants are replaced by **algae**. They are not classified as plants because they don't have separate roots, stems, and leaves.

Algae include large seaweeds and microscopic phytoplankton like those at the bottom of page 53. Like plants, algae are important to other living things. They provide nutrients and oxygen to support life in the oceans.

Naming plants

In 1735, a Swedish plant expert called Carl Linnaeus invented the system we use to name species. He chose Latin names because that was the international language of science at the time.

Linnaeus was good at getting publicity. His students enjoyed his lessons. He organised popular plant-finding trips and he wrote lots of books. Everyone heard about his system so they started to use it themselves.

Scientists need to be good communicators to get people to understand their ideas.

↑ Algae have no separate roots, stems and leaves.

1. These drawings show four groups of plants. Copy each drawing and name the group it represents.
2. Add labels to each drawing to show the main features of each group.
3. Which groups should these newly discovered species be placed in?

    Plant A is 1 metre tall, has feathery leaves and it never flowers.

    Plant B is very tall and produces seeds in cones.

    Plant C is very low growing and can only survive in damp places.

    Plant D is tall and thin and its flowers are almost invisible.

4. Spot the odd one out in each of these sets and explain your choice:

    a grass, fern, flowering tree

    b moss, conifer, flowering tree

    c moss, grass, fern

    d fern, conifer, moss

    e algae, ferns, flowering plants

5. Surtsey is a volcanic island near Iceland. Soon after it formed, some plants began to grow on the bare rock. The plants were only a few millimetres tall and had no roots. What were they?
6. Ten years later gulls began to nest on Surtsey and the number of flowering plants growing there suddenly increased. Suggest why.

- Plants include mosses, ferns, conifers, and flowering plants.
- Flowering plants grow from seeds produced in flowers.
- Conifers grow from seeds produced in cones.
- Mosses and ferns reproduce using spores.
- Mosses have no roots or veins so they can't grow tall.

Review 5.7

1. The chart below shows the eye colours of a group of students.

   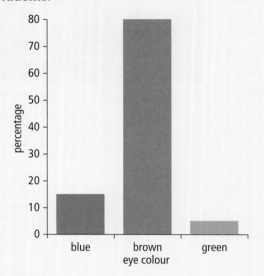

   a. Which eye colour is least common? [1]
   b. Name another human feature that shows discontinuous variation like this. [1]

2. The chart below shows the hand spans of a group of students.

   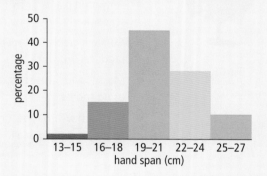

   a. What percentage of students had hand spans of 22 cm or more? [1]
   b. Name two other human features that show continuous variation like this. [2]

3. The table below shows the variation in the colour of a population of beetles.

Colour	Percentage of beetles
black	30
dark brown	60
pale brown	10

   Display the data on a frequency chart. [2]

4. The table below shows how the lengths of a species of beetle varied.

Length (mm)	Percentage of beetles
10–11	8
12–13	22
14–15	45
16–17	20
18–19	5

   Display the data on a frequency chart. [2]

5. The scientific names of two bears are given below:

   polar bear – *Ursus maritimus*
   American black bear – *Ursus americanus*

   a. Explain why their second names are different. [1]
   b. Suggest why they share the same first name. [1]
   c. Explain why species need scientific names. [1]

6. Cheetahs can run faster than any other animal. They are found in many parts of Africa. Explain how scientists could check that the cheetahs in separate groups belong to the same species. [3]

7. Zedonks are born when a female donkey breeds with a male zebra.

   a. What name is given to animals like the zedonk? [1]
   b. Explain why two zedonks will never produce offspring. [2]

8. The diagram shows a selection of animals.

   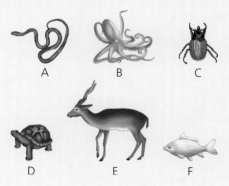

a Choose two that are invertebrates. [2]
b What is the same about the vertebrates? [1]
c Choose two vertebrates that are reptiles. [2]
d Describe two features that reptiles share. [2]

9 Name the vertebrate group each of these animals belong to.

A   B   C   D

10 The duck-billed platypus spends a lot of time in water. It eats small invertebrates which it crushes with its bill. Females lay eggs and keep them warm until they hatch.

a Which of its features shows that it is a mammal? [1]
b Which feature does it share with reptiles and birds? [2]

11 Rats are mammals.

Write down three ways in which mammals are different from other vertebrates. [3]

12 Three different invertebrate groups contain worms:

Group	Characteristics
roundworms	no segments, round
annelids	segments
flatworms	no segments, flat

Which group does this animal belong to? [2]

13 Explain why a fly is classified as an insect but a spider is classified as an arachnid. [2]

fly   spider

14 This animal feeds on dung. Use its features to decide which of the following groups it belongs to.

a Vertebrate or invertebrate? [1]
b Mollusc or arthropod? [1]
c Insect or crustacean? [1]

15 Which of these plants produces seeds? [1]

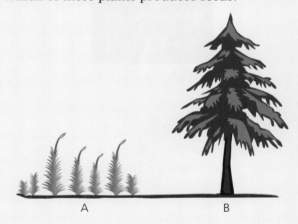

A   B

16 Choose the correct plant group for each of the following:

a A small, non-flowering plant with thin leaves. It grows in clusters in dark, damp conditions. [1]
b Large, non-flowering plants with stems, leaves, and roots. The leaves have fronds that have spores on their undersides. [1]
c Woody, non-flowering plants with needle-like leaves. [1]

Review Stage 7

1 The parts of this plant cell are labelled A–F.

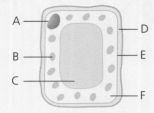

   a Give the correct letter for each of these parts:
      i   cell wall   [1]
      ii   cytoplasm   [1]
      iii vacuole   [1]
      iv   membrane.   [1]
   b Which part makes plants look green?   [1]
   c Which three parts are found in animal cells? [3]

2 The image shows a knee X-ray.

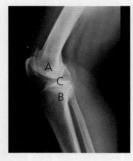

   a What are A and B made of?   [1]
   b Describe their function.   [1]
   c What name is given to the gap C?   [1]
   d The tissues that move A and B don't show on the X-ray. What are they called?   [1]

3 Water hyacinth grows on the surface of lakes. Weevils slow down its growth by eating its leaves. The weevils themselves are attacked by nematode worms.

weevil

   a Draw the food chain that links these organisms.   [1]
   b Which organism is a producer?   [1]
   c Which is a herbivore?   [1]

4 These animals are both found in many countries.

snake   frog

   a These animals are both vertebrates. What feature do all vertebrates have?   [1]
   b Identify the vertebrate group each animal belongs to.   [2]
   c Which animal has dry, scaly skin?   [1]

5 You need to find out how much acid it takes to damage seedlings. You have four pots of seedlings and a bottle of dilute acid.

   a What will you vary?   [1]
   b What will you keep the same?   [1]
   c Suggest one way of comparing how much the seedlings are damaged.   [1]

6 The diagrams show some specialised cells.

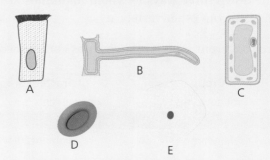

   a Which cell provides a large surface area to increase the uptake of minerals?   [1]
   b Which is small, flexible, and good at carrying oxygen?   [1]
   c Which specialises in making food?   [1]

7 The table gives data about two fuels that can be used in cars.

Fuel	Source	Carbon dioxide released?	Sulfur dioxide released?
petrol	oil	yes	yes
bioethanol	sugar cane	yes	no

  a Which fuel contributes most to acid rain? [1]
  b Which fuel is renewable? [1]
  c Name two other renewable energy sources. [1]

8 These plants grow on the branches of tall trees high up in the South American rainforest.

  a Which part of a plant produces seeds? [1]
  b Suggest why these plants grow best on the highest branches. [1]
  c The plant has short roots but no root hairs. In other plants, what do root hairs collect from soil? [2]

9 The diagram shows some muscles and bones in an arm.

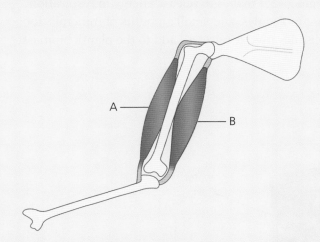

  a Which muscle straightens your arm when it contracts? [1]
  b Explain why there are two muscles in your upper arm. [1]

10 Amoeba is a micro-organism found in pond water. When it senses food it moves towards it, wraps its cytoplasm around it, and digests it. It uses some of the glucose in the food for respiration.

List three **other** life processes the amoeba must carry out. [3]

11 Giraffes are found in several parts of Africa.
  a What word is a place where an animal lives? [1]
  b Giraffes have different coat patterns and heights. What word describes differences between members of a species? [1]
  c Giraffes use their long necks to reach the leaves on acacia trees. What name is given to features like this that help organisms survive? [1]

12 Milk was left in the fridge for 3 days.

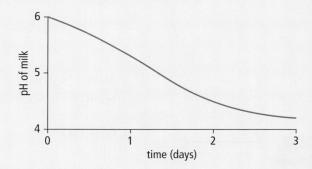

  a Describe how its pH changed. [1]
  b What caused the pH to change? [1]
  c What difference would you expect if the milk was left in a warmer place? [1]

13 A hospital compared the number of live bacteria left after washing with soap or hand sanitiser.

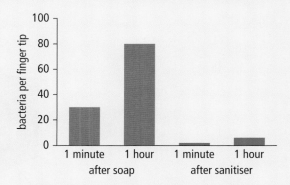

State two conclusions that can be drawn from these results. [2]

6.1 Why we need plants

Objective
- Describe the importance of plants to life on Earth

Biomass

All the wood in these tree trunks is **biomass**, which is the material living things are made of. This biomass was made from air and water. How is that possible?

Plant cells use chemical reactions to build biomass. To do this they need two small molecules – carbon dioxide and water – and energy from the Sun.

The solar energy enables plants to build larger molecules such as glucose from the atoms in carbon dioxide and water. The reaction also releases oxygen. This chemical reaction is **photosynthesis**. It sustains life on Earth.

↑ These massive tree trunks are made from air and water.

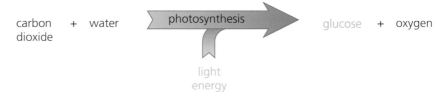

carbon dioxide + water →[photosynthesis, light energy]→ glucose + oxygen

All the food we eat comes from the glucose plants make. All the oxygen we breathe comes from the oxygen they release.

Energy

Glucose molecules contain stored energy. Cells can release energy from glucose using respiration. Photosynthesis and respiration are the reverse of each other. Photosynthesis stores energy, and respiration releases it.

A plant uses about half the glucose it makes to release energy in respiration. It uses the rest of the glucose molecules, and small amounts of minerals, for growth and repair. The molecules used for growth add to the plant's biomass.

↑ Plants use some of the glucose they make for respiration and the rest for growth.

glucose → (respiration) → carbon dioxide + water, chemical energy released
glucose → (growth) → extra biomass

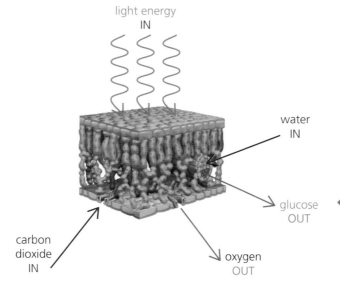

← This section through a leaf shows the different types of cell. Leaves are designed to bring water and carbon dioxide to their cells for photosynthesis.

Leaves

Most photosynthesis takes place in the tall thin **palisade** cells near the top of a leaf. You can take a closer look at them on page 43. They have plenty of **chloroplasts** to absorb light energy.

Veins bring water and minerals from the roots up the stem to the leaves. The water and minerals travel along hollow tubes called **xylem** (see pages 88 and 89).

Carbon dioxide diffuses into the leaf from the air through tiny pores called **stomata**. The **spongy mesophyll** layer at the bottom of the leaf makes it easy for gases to circulate. Any oxygen that the cells don't need diffuses out through the stomata.

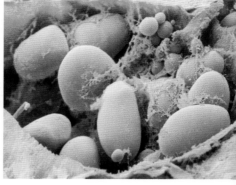

↑ This SEM image shows the starch grains inside a potato cell – magnified 640 times.

Starch

Plant cells need to store some glucose so their cells can respire at night when photosynthesis stops. Small, soluble molecules such as glucose can't be stored in cells. Instead, they are joined together to make giant molecules of starch.

Leaves can store enough starch to last for 2 or 3 days. Starch is easy to detect because it makes brown iodine solution turn a dark blue-black colour.

Some of the glucose made by leaf cells is sent to cells that can't make their own glucose. It travels down to the roots in phloem tubes inside veins.

Oxygen

Instead of taking carbon dioxide for photosynthesis from the air, pondweed takes it from water.

You can see when photosynthesis is taking place in pondweed. The spare oxygen the plant releases forms bubbles in the water. This is useful if you want to measure how fast the photosynthesis reaction is going. You can count the bubbles or measure the volume of oxygen produced.

↑ Photosynthesis makes bubbles of oxygen escape from pondweed.

Q

1. Give two reasons why we need plants.
2. Draw an outline of a leaf. Then add arrows to show which molecules enter it and which leave when the plant is photosynthesising.
3. How do plants use the glucose they make?
4. How do they store the glucose that they can't use straight away?
5. Plants and animals both use respiration to release energy. Where did this energy originally come from?
6. The diagram on the right shows a section through a leaf. Copy the diagram and label these parts:
   a. one of the stomata that lets gases diffuse in and out of the leaf
   b. the spongy mesophyll that lets gases move between the cells
   c. the palisade cells where most photosynthesis takes place
   d. xylem tissue that carries water up from the roots
   e. phloem tissue that carries glucose down to the roots.

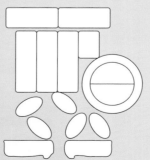

!

- Plants use photosynthesis to produce glucose and oxygen.
- Some glucose is used for respiration, and some is stored as starch.
- Glucose and minerals are used to build biomass.
- Plant biomass provides food for animals.

Enquiry 6.2

Asking scientific questions

Planning an investigation

Rahul needs to plan an investigation. He wants to find out how to make plants grow better. Before he can start, he needs to turn this idea into a **scientific question**.

Scientific questions can be answered by collecting evidence. 'Which plant is best?' is not a scientific question because it is not clear what 'best' means. A good scientific question involves two variables: one that can be changed to answer the question, and one that can be measured to see what effect that change has.

Rahul must use what he knows about plants to think of something that could affect their growth. Then he needs to find a variable he can measure to show how well they are growing.

Choosing variables

Rahul knows that light, carbon dioxide, and water are needed for photosynthesis, and that plants also take in minerals. But which of these factors affect how well a plant grows?

The light intensity could be important because light supplies the energy plants need. This is the variable he will test. Now he needs a variable he can measure.

Rahul could measure how fast plants increase their biomass. This would take a long time though. Plants grow slowly. They also have a lot of water in them. He would need to dry them out to see how much their biomass increased. He needs a measurement he can collect more easily.

Plants need to use the nutrients they make in order to grow. When they make nutrients faster, they usually grow faster. Rahul could give plants different amounts of light and measure how fast photosynthesis takes place (the **rate** of photosynthesis).

Objectives

- Understand what makes a question scientific
- Be able to develop a scientific question that can be investigated

⬆ Any of these variables could affect how well a plant grows.
- light intensity
- carbon dioxide concentration
- water supply
- mineral supply

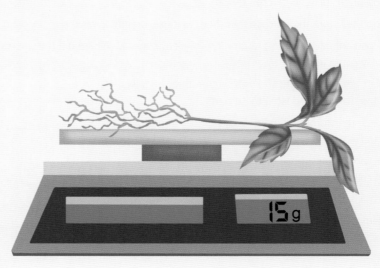

➡ Plants contain a lot of water. To measure their biomass accurately they need to be dried. The biomass of this plant is much less than 15 grams.

Rahul could use pondweed for his investigation. He could compare its rate of photosynthesis in different light intensities by measuring the volume of oxygen it releases per day.

The question Rahul plans to answer is:

Does light intensity affect the volume of oxygen that pondweed releases each day?

Rahul predicts that the light intensity will make a big difference. He thinks that if more light energy shines on the plant's leaves, they should absorb light energy faster. That should make the photosynthesis reaction faster.

Collecting results

Rahul traps pondweed under a funnel and collects the oxygen it produces in a measuring cylinder. He makes five identical sets of apparatus. He puts each set a different distance from a sunny window to change the light intensity. He will measure the volume of oxygen in each set of apparatus at the end of the day.

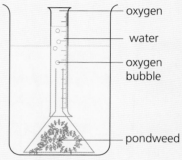

↑ The greater the rate of photosynthesis, the faster oxygen collects.

Presenting the results

A line graph makes it easier to identify any trends or patterns in the results. Rahul's graph shows that his prediction was correct. The light intensity does affect the volume of oxygen produced. The closer the pondweed was to the window, the more oxygen it produced.

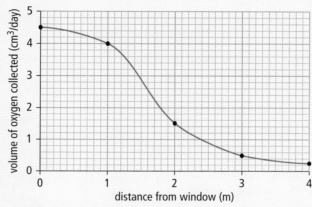

↑ The volume of oxygen collected per day falls as the distance from the window increases, because the light intensity gets less further away from the window.

Q

1. Give two reasons why this is a scientific question: 'Does the temperature of the water affect the time it takes pondweed to release 1 cm³ of oxygen?'
2. Rahul measured the volume of oxygen released in a certain time. Why is this a good measure of the rate of photosynthesis?
3. Explain why the rate of photosynthesis shows how fast a plant can grow.
4. Suggest two variables Rahul needed to control in order to compare the light intensities fairly.
5. Why is it useful to plot a graph of the results?

- A scientific question can be answered by doing an investigation.
- It involves a variable that can be changed or compared and a variable that can be observed or measured.

6.3 Water and minerals

Objective
- Describe how water and minerals are absorbed by roots and transported to leaves

Why plants need water

Plants need water because their cells use water for photosynthesis, and water evaporates from their leaves. If a plant doesn't get enough water, it **wilts**. That means its leaves and stem droop because their cells cannot support themselves.

Normally each cell's vacuole pushes against its cell walls (see page 43). This keeps the cell **turgid**, which makes it firm and rigid.

⬆ This plant has wilted. Its cells cannot support themselves without water.

If a plant can't take in enough water, each cell's vacuole shrinks. Then there's nothing to press the cytoplasm against the cell wall. The cell becomes **flaccid**. Flaccid cells are floppy, so they can't hold a plant up.

Water also carries minerals to cells. The minerals are stored in the vacuole until the cell needs them.

 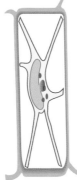

turgid cell   flaccid cell in a plant that is short of water

⬆ Flaccid cells are floppy so they cannot stand upright.

How do water and minerals enter and leave the plant?

The flow of water through a plant is called **transpiration**.

It begins when water evaporates and escapes through gaps called **stomata** on the undersides of the leaves. This pulls more water up through **xylem** vessels to replace it.

Xylem vessels are hollow tubes. They run through the stem connecting the veins in the leaves to the roots. The roots maintain the plant's water supply by absorbing it from the soil. Root hair cells (see page 45) give the root a very large surface area to make this easier. They also use some of the energy they get from respiration to take minerals in from the soil.

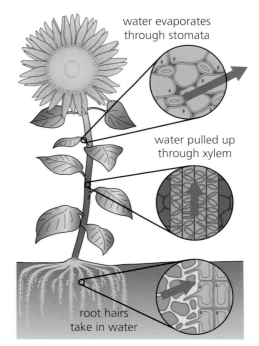

⬆ Evaporation from the leaves makes water flow continuously from the roots to the leaves.

Plants

Controlling the flow of water

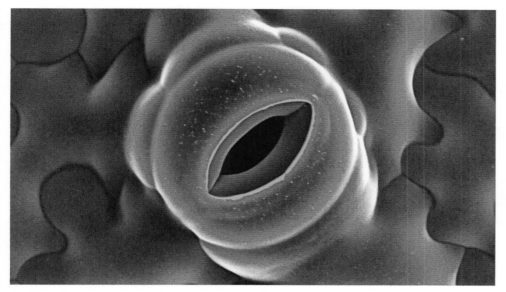

↑ SEM of guard cells around a stomata – magnified 960 times.

Stomata can be opened and closed. Each **stoma** is surrounded by two specialised cells called **guard cells**. At night, leaves don't need to take carbon dioxide in for photosynthesis, so the stomata close. They can also close during the day if a leaf is short of water.

Guard cells have specialised cell walls that make them push each other apart when they are turgid. When they lose water and become flaccid, the gap between them closes.

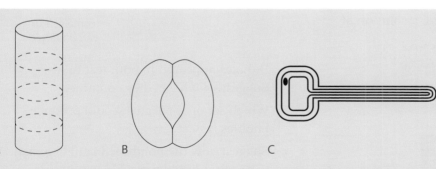

- Water and minerals are absorbed from soil by plant roots. They travel up to the leaves in xylem vessels.
- Water evaporates and escapes through stomata.

1. The diagram shows the shapes of three specialised cells involved in transpiration. Name each cell and explain how its shape suits its function.

2. The diagram on the right shows a leaf trapped between two slides. Under each slide is a piece of cobalt chloride paper. This paper turns pink when water touches it. Explain why the paper on the bottom of the leaf turns pink before the paper on the top of the leaf.

3. Describe the route a water molecule takes from the soil to the atmosphere if it is not used for photosynthesis. Mention any specialised cells it passes through on its way.

4. Draw a plant cell that has lost a lot of water. Explain how a plant is affected if this happens to a lot of its cells.

5. Explain why plants need small amounts of minerals. Describe how they obtain these minerals.

Review 6.4

1. *Convolvulus arvensis* is a flowering plant. Its stems cannot stand upright so it spirals around other plants.

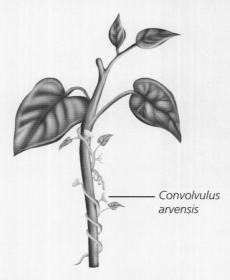
Convolvulus arvensis

  a. Suggest how climbing to the top of other plants helps *Convolvulus* survive. [1]

  b. If the stem of the *Convolvulus* plant is cut, the leaves above the cut die. Suggest why. [1]

2. Hydrogencarbonate indicator is orange. When carbon dioxide is added or removed, a solution of the indicator changes colour.

Carbon dioxide	Colour of indicator
added	yellow
removed	purple

orange hydrogencarbonate indicator

  a. When pondweed was left in a tube of the indicator for a day, the solution went purple. Explain why. [2]

  b. A second tube of pondweed and indicator left in the dark went yellow. Explain why. [1]

3. Three identical beakers of pondweed were left in darkness, dim light, or intense light for an hour. Then the oxygen in each beaker was measured using an oxygen sensor.

Lighting	Dissolved oxygen (mg/dm³)
darkness	5
dim light	14
intense light	20

  a. Describe how the amount of oxygen in the water was affected by the light intensity. [1]

  b. How would the light intensity affect the amount of carbon dioxide in the water? [1]

4. Pondweed was placed in two identical sets of apparatus. A chemical was added to the water in apparatus B to increase the carbon dioxide concentration in the water.

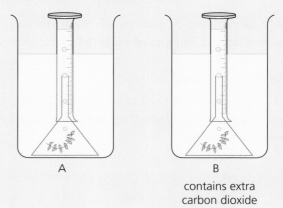
A   B
contains extra carbon dioxide

  a. Bubbles appeared in both test tubes. Suggest which gas they contained. [1]

  b. Which set of apparatus would produce bubbles fastest? [1]

  c. Suggest how the pondweed could be made to produce bubbles even faster. [1]

5. When oil is spread on leaves, it stops them losing water. Two identical plants had different leaf surfaces covered in oil. They were weighed at the start and after 1 day to see how much water they lost.

a. oil on top surface of the leaves   b. oil on bottom surface of the leaves

a Suggest which plant will lose more water and give a reason for your answer. [2]

b Explain why their pots were covered in plastic bags. [1]

6 The diagram shows a microscope image of the underside of a leaf.

a Identify the four structures shown. [1]

b Explain why leaves need these structures. [1]

c In wetter climates plants usually have more stomata and grow faster. Suggest why. [2]

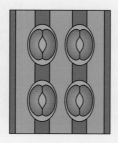

7 The diagram shows a cell from near the top of a leaf.

a What type of cell is this? [1]

b Why is it useful for these cells to be near the top of a leaf? [1]

c Under higher magnification, tiny round starch grains can be seen the cell's cytoplasm. Explain what cells use these starch grains for. [2]

d How would the number of starch grains change if the plant was left in the dark for a few days? Explain your answer. [2]

8 Identical seedlings were placed in pure water or water that had been mixed with soil.

A. pure water   B. water mixed with soil

The table compares their appearance after 2 weeks.

Plant	Height of plant (cm)	Observations
A	9	yellow leaves no flowers
B	15	green leaves flower buds

Suggest a reason for the differences listed in the table. [2]

9 The diagram shows sections taken from two different stems.

untreated stem   stem placed in ink for a day

a Identify the part of the stem that turned black in ink. [1]

b Describe the job this tissue does. [1]

10 A plant was left in the dark for 2 days to use up its stores of starch. Then two of its leaves were treated as shown. At the end of the day leaves A and B were tested with iodine.

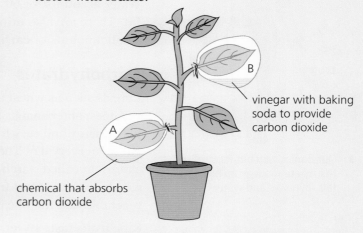

a Suggest what scientific question this investigation is designed to answer. [1]

b Predict which leaf would turn blue-black with iodine. [1]

11 If a plastic bag is tied around a plant, drops of water appear inside the bag.

a Explain where the water is coming from. [1]

b What name is given to the flow of water through a plant? [1]

12 Do plants have fewer stomata in places with lower annual rainfall?

a Explain why this is a scientific question. [2]

b Predict what you might find if you collected evidence to answer the question. Give reasons. [2]

7.1 Food

Objectives
- List the nutrients in food
- Understand why each nutrient is needed

Why eat?

Do you feel healthy and full of energy? If so you are probably getting the right amount of food. The **nutrients** in food have three roles. They provide materials for growth and repair, energy to keep your cells alive, and vital elements and compounds to maintain the chemical reactions in your cells.

↑ Eating the right food can keep you healthy and energetic.

Your body needs the types of nutrients shown in the diagram. Water accounts for 60% of your body mass. The rest is mostly **protein** and **fat**. There are also **minerals** in your bones and teeth, and your body contains small amounts of **carbohydrate**.

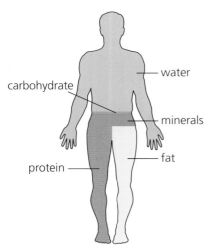

↑ Our bodies vary but most of our tissues are made from proteins, fats, minerals and water.

Carbohydrates

Cereals such as wheat and rice, beans, potatoes, and bananas are packed with energy and easy to grow. People eat them every day. They all contain a nutrient called starch. Starch is a carbohydrate. Most fruits and vegetables are full of carbohydrates.

Ripe fruits taste sweet. They contain sugars such as glucose. These are carbohydrates too. You can see what makes sugars different from starch by turning back to page 85. Starch is made of lots of glucose molecules joined together.

↑ All these foods contain carbohydrates.

Foods like chocolate and cakes contain large quantities of sugar. They give you a quick burst of energy. Simple sugars such as glucose don't need to be digested – they pass straight from your small intestine into your blood unchanged.

Proteins

Meat is muscle so it contains a lot of protein. Protein is also found in fish, eggs, nuts, beans, and foods made from milk.

Some of your proteins build structures like muscle fibres. Other proteins organise the chemical reactions in your cells. Your genes include recipes for making thousands of different proteins. Proteins are more complicated than starch. They are made from 20 different components called **amino acids**.

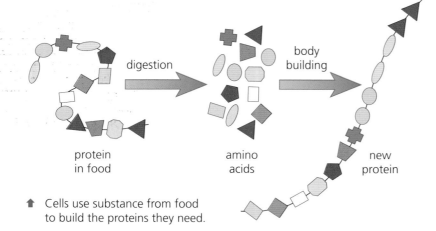

↑ Cells use substance from food to build the proteins they need.

Fats

Fats are an important source of energy, like carbohydrates. They are also used for building cells. The membrane of every cell is made of fat. Nerves use extra fat as insulation. Fat cells (page 44) protect vital organs such as your heart, and reduce heat loss from your skin. They can also expand to store excess food for later.

Fats are found in meat, fish, milk products, and many seeds and nuts. They are built from two types of molecule – glycerol and fatty acids. Each glycerol molecule is joined to three fatty acids. There are many different fatty acids and they determine the properties and energy content of each type of fat.

↑ Different fats and oils contain different fatty acids.

Vitamins and minerals

You need **vitamins** and minerals in much smaller amounts than other nutrients. Vitamins help chemical reactions take place in your cells. Different foods contain different vitamins. The vitamins in fish and dairy products dissolve in fats, so you can store them until you need them. The ones in fruit and vegetables can't be stored so you need to eat them regularly.

There are minerals in your bones and teeth, in all your cells and in all your body fluids. Some minerals provide strength and others help cells to function properly.

Fibre

Beans and lentils are good sources of dietary **fibre**.

Fibre is an essential part of your diet but it isn't a nutrient. Fibre is made of large molecules that your body cannot digest – mostly plant cell walls. The fibre in food helps it to pass through your digestive system quickly, so it prevents **constipation**.

↑ Beans and lentils are good sources of dietary fibre.

Q

1. List the three reasons why your body needs to take in nutrients.
2. Which two types of nutrient are needed to build new cells?
3. Name the two nutrients that provide energy.
4. Which nutrients help maintain the chemical reactions in your cells?
5. Proteins, fats, and carbohydrates are large molecules made by joining smaller molecules together. Draw diagrams to explain their structures.
6. Draw a table and list the five types of nutrient your body needs. For each nutrient, give an example of a food that contains it.

Food contains five types of nutrient:

- carbohydrates, for energy
- proteins, for growth and repair
- fat, for insulation and making cell membranes
- vitamins, to aid chemical reactions in cells
- minerals, for growth and to aid the activity of cells

It also contains fibre, which prevents constipation.

Enquiry 7.2

Managing variables

Objectives
- Understand how scientists can measure the energy content of food
- Work out what variables must be changed, controlled, and measured

Which foods give us most energy?

To find out which foods give us most energy, we need to compare foods and take careful measurements. But what could we measure?

One idea is to burn the foods. When foods burn they release most of their energy as heat. We can capture this heat by using it to warm up water. The more energy the food releases, the more the water's temperature rises.

↑ Foods release energy as heat when they burn.

Taking accurate measurements

To collect accurate results we need to measure the temperature of the water twice – before the food starts burning, and after it has all burnt. The difference between these values will show us the temperature rise each food causes.

Comparing foods fairly

To compare foods fairly we need to ask another question:

What else could affect the water's temperature rise besides the type of food we are testing?

One thing that will affect the temperature rise is the volume of water we use. The less water there is in the test tube, the less energy it will take to heat it. We can control this variable by using the same volume of water for every food we test.

We need to choose a sensible volume. If we don't use enough water, it could boil away before we collect our results. If we use too much, its temperature rise might be too small to measure accurately. 20 cm³ seems a sensible volume.

Before we start collecting measurements we need to choose values for every other variable that needs to be controlled. Then it's a good idea to check that the method works.

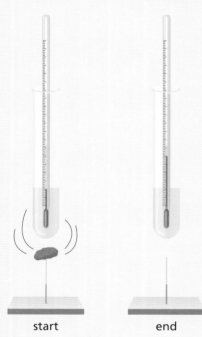

start   end

↑ The bigger the temperature rise, the more energy the food contains.

↑ A measuring cylinder can be used to measure volumes accurately.

Checking the method

Eniola was planning to compare the energy in foods. To check her method, she tested pieces of the same food with masses of 0.5 g and 1 g. She used a flame to set light to them. Then she used each piece of burning food to raise the temperature of a tube of water.

Food	Start temperature (°C)	End temperature (°C)	Temperature rise (°C)
bread (0.5 g)	20	50	30
bread (1.0 g)	20	80	60

The burning food made a lot of smoke. She had to be careful not to burn herself because the apparatus got very hot. Eniola was pleased with her results though. The method seemed to work. When she doubled the mass of bread, the temperature rise doubled.

A better method

Although Eniola's method worked, she noticed things she could improve.

It took a long time to get pieces of food with an exact mass. She had to keep cutting bits off them. Her investigation would be quicker if she didn't have to do that. There was another problem. The food stopped burning before it was all gone.

Eniola changed her method slightly to overcome these problems. She measured the mass of each food at the start and at the end. Then she worked out the difference to see how much food had burnt.

start · end

 The difference in mass shows how much food has burnt.

Eniola realised she could test any mass of food that fitted under the tube of water. Then she could calculate the temperature rise 1 g of food would cause.

Temperature rise per gram = temperature rise ÷ mass of food used
(°C per gram)　　　　　　　(°C)　　　　　　(g)

Q

1. Describe how you could compare the amounts of energy in two foods. Explain what you would need to measure and how you would know which food has more energy.
2. List all the variables that need to be controlled in this experiment and suggest suitable values to use.
3. Suggest why Eniola was pleased with her results.
4. Suppose the water had boiled when Eniola tested 1 g of bread. Suggest two changes she could have made to her method.
5. Eniola's results turned out lower than she expected. Suggest two possible reasons why she didn't collect all the energy in the foods she tested.

To plan an investigation you need to:
- decide what to compare and what to measure
- identify variables that need to be controlled
- choose a suitable value for each control variable
- test that your chosen values work.

7.3 A balanced diet

Objective
- Understand what a balanced diet is

What is a balanced diet?

To stay healthy you need a balanced diet. This is a diet that contains the right amounts of all the nutrients your cells need to function properly. To get a balanced diet, you have to eat a wide variety of foods.

Nutritional requirements vary from country to country. They are higher in colder places and where people get more exercise. They are also different for males and females and different age groups. But some recommendations are the same for everyone. Your diet needs to replace the energy you use and give you a suitable mix of proteins, fats, and carbohydrates.

Proteins

Scientists estimate that 10–13-year-olds need 34 g of protein per day. It should come from a variety of different foods. Each type of protein is made from different amino acids. Our cells can make some of them, but there are others we can only get from food.

↑ Soya beans supply more protein than any other crop per square metre of land.

Meat, eggs, soya beans, and dairy products supply all the amino acids we need in the right proportions. Most plant products need to be combined to supply a full set of amino acids.

Fat

Fats should make up 20–30% of the food you eat. They supply more than twice as much energy per gram as carbohydrates. They also contain some essential vitamins.

There are two main types of fat, saturated and unsaturated. Processed foods are often high in saturated fat. You should minimise the amount of these you eat. They increase your risk of heart disease. Fish, nuts, and vegetable oils contain unsaturated fats, which are better for your health.

Fats contain fatty acids. Two types of fatty acid are essential. Your cells need them but can't make them. They are called omega-3 and omega-6 fatty acids.

↑ Fish are a good source of essential fatty acids.

Carbohydrate

You should get about 60% of your energy from carbohydrates. As many of these as possible should come from fruits and vegetables because these also give you plenty of vitamins, minerals, and fibre. The rest should come from starchy foods like rice, cassava, and bread.

Malnutrition

Many people don't have a balanced diet. They suffer from **malnutrition**. Some people don't get enough energy and nutrients to function normally. They can't grow to their full height and they suffer health problems.

Others eat too much. Their bodies store this extra energy as fat. The excess fat makes them more likely to suffer from heart disease or diabetes when they are older.

Sweet food and fizzy drinks cause another problem. Their sugar coats our teeth and gums and encourages microbes to grow there. Some of these microbes convert the sugar to acid which causes tooth decay.

← Vegetables provide carbohydrates, vitamins, and minerals.

Calculating how much food you need

On average, students need to eat around 2000 kilocalories per day, which is about 8000 kilojoules (kJ) per day. Each gram of carbohydrate supplies 16 kJ, and 60% of your energy should come from carbohydrates, so we can calculate the recommended mass of carbohydrate you should eat each day.

Mass of carbohydrate recommended = $\frac{60}{100}$ × energy needed ÷ energy supplied
(g per day)   (kJ per day)   (kJ per g)

$$= \frac{60}{100} \times 8000 \div 16 = 300 \text{ g}$$

1. Which type of nutrient should you eat most of?
2. Explain why we should get our proteins from a variety of different foods.
3. Rice and beans are often eaten together. Together they supply all the amino acids we need. Explain what amino acids are and why we need them.
4. Eating too many fats can damage your health. Give two reasons why it is essential to eat some fats.
5. Give two reasons why eating too many sweets can damage your health.
6. **Extension:** fat supplies 37 kJ per g. How much fat would you need to eat to get 40% of your daily energy requirements?

- A balanced diet contains every essential nutrient, in the correct proportions.
- The body stores energy foods that it can't can't use straight away as fat.
- Eating sugar can lead to tooth decay.

7.4 Deficiencies

Objectives
- Recall some of the main roles of specific vitamins and minerals
- Discuss the importance of collecting evidence, developing explanations, and using creative thinking

Scurvy

Between 1500 and 1800, being a sailor was a hazardous occupation. After 12 weeks at sea most sailors developed **scurvy**. Their gums bled, their legs swelled up, and they had no energy at all. Many died.

If sailors had fresh orange or lemon juice they survived, but these were expensive and hard to store. Two popular ideas were that anything sour would cure scurvy, and that lemon juice would still work if it was boiled to make it last longer. Unfortunately neither idea worked. Sailors continued to die.

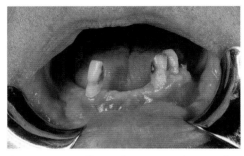
↑ A person suffering from scurvy has swollen gums and loses teeth.

Beri-beri

In 1910, Casimir Funk investigated a disease called **beri-beri**. It causes partial paralysis and mental confusion. Other scientists had shown that it wasn't caused by microbes. People got the disease if they ate mainly white rice instead of brown rice. White rice is just brown rice with its skin taken off. Funk used creative thinking to explain this observation: *Rice skins must contain an essential nutrient.* He needed to collect evidence to test his explanation.

Pigeons get a similar disease, so Funk fed sick pigeons different chemicals from rice skins. One of the chemicals cured them, and he called it vitamin B1.

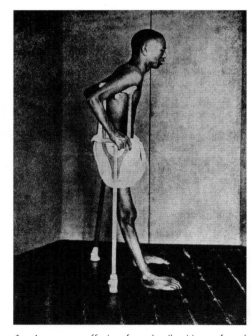
↑ A person suffering from beriberi is confused and partially paralysed.

Vitamins

Funk predicted that vitamins would cure other diseases. This idea spread and other scientists began to find vitamins. Vitamin A in egg yolk cured **night blindness** and vitamin D in cod liver oil prevented **rickets**.

Eventually, in the 1920s, vitamin C was found in oranges and lemons. This essential vitamin is destroyed by boiling.

↑ A child suffering from rickets had soft, bent bones.

Deficiency diseases

Scurvy, beri-beri, night blindness, and rickets are all **deficiency diseases**. Diseases like this occur when people don't get enough of an essential nutrient. In the past these diseases were common in places like prisons where people were fed the cheapest possible food.

Vitamin D is made when skin is exposed to sunlight, so it should never be deficient in people who spend time outside.

In some parts of the world people can't get enough proteins. This causes **kwashiorkor**. Its symptoms include muscle shrinkage and a swollen belly.

Deficiency diseases can also be caused by not getting enough minerals. A shortage of calcium causes rickets, just like vitamin D deficiency. A lack of iron leads to **anaemia**. This causes tiredness because we need iron to build the blood cells that carry oxygen. People with anaemia also have painful sores in their mouths and weak nails.

Many other minerals are needed in smaller amounts. Everyone should get enough of all the vitamins and minerals if they eat a balanced diet.

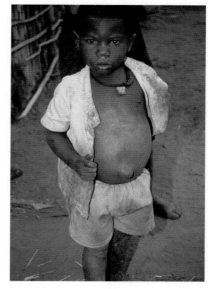

↑ This child has kwashiorkor caused by not eating enough protein.

Too much of a good thing?

At first scientists assumed that taking extra vitamins would always be good for you. Then they began to find evidence that large quantities could be harmful.

Early Arctic explorers were poisoned when they ate polar bear liver. It contained huge quantities of vitamin A. Other animals' livers contain less vitamin A, but pregnant women are advised not to eat liver at all. Small amounts of vitamin A help a baby's organs develop, but too much can be harmful.

Other vitamins and minerals can be harmful in large quantities. Safety committees limit the amounts that can be put into vitamin tablets.

↑ Pregnant women are advised not to eat liver.

Q

1. Explain what a deficiency disease is.
2. Draw a table that shows:
   - some nutrient deficiency diseases
   - the main symptoms of each disease
   - the missing nutrient which causes each disease.
3. Suggest why early sailors didn't get enough vitamin C on long voyages.
4. Explain why lemon juice cured them but boiled lemon juice didn't work.
5. Explain why vitamin D deficiency is more common in countries further away from the equator.
6. How did Casimir Funk collect evidence to support his ideas?

- Scientists use creative thinking to develop new explanations
- Their ideas need to be tested to check there is evidence to support them
- When people don't get enough of an essential nutrient they may develop a deficiency disease.
- Deficiency diseases can often be cured by supplying the missing nutrient.

Extension 7.5

Objective
- Understand why people don't always have balanced diets

Choosing foods

Favourite foods

Think of your favourite food. Is it very sweet or full of fat? Nearly everybody likes sweet, fatty foods. They make us relax.

Thousands of years ago, sugary foods were hard to find. Our ancestors had to search for them. Sugary foods contain lots of energy, so they helped to keep our ancestors alive. We inherited their love of sugar, but we can easily eat too much. Sugary foods are much easier to find now.

Bitter foods have the opposite effect. They make you want to spit them out. Most poisonous plants taste bitter, so that's useful.

Overeating

In most countries average body masses are rising. All the food we don't use is stored as fat. If someone's body mass is much higher than the recommended level, we say they are **obese**. In the worst affected countries, more than 30% of adults are obese, and the number continues to rise. Scientists blame a change in people's eating habits.

In many parts of the world families are eating fewer fresh foods. They rely more on takeaways and processed foods from supermarkets – especially for teenagers. These contain more sugar, salt, and saturated fats than fresh foods, which makes them very tasty. At the same time, we have more energy-saving machines in our homes and more cars, so we get less exercise.

Having a high body mass causes lots of long-term health problems including diabetes, cancer, high blood pressure, and heart disease. Health authorities are worried because these could cost billions to treat in the future.

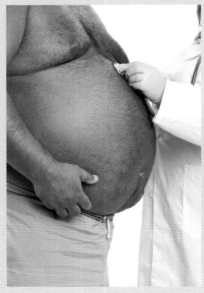

↑ Obesity causes numerous health problems and shortens lives.

Solutions

Scientists argue that educating people about balanced diets isn't enough. The only way to prevent obesity is to make fatty and sugary foods more expensive and get people to take more exercise. That means changing the design of cities to make walking or cycling more attractive than owning a car.

Until that happens they want to stop TV companies advertising sugary foods and prevent them being sold in schools.

↑ Some new offices have stairs and ramps instead of lifts to give workers more exercise.

Diet

Tackling deficiency diseases

Even in countries where obesity is common, people can still suffer from nutritional deficiencies.

Wheat is one of the world's main cereal crops. Wheat seeds are used to make flour. Modern wheat plants have bigger seeds, but they contain fewer vitamins and minerals than older crops.

In most countries, common foods like bread and cereals are **fortified**, which means that extra nutrients are added to them. This makes it easy for people to get more vitamins and minerals without changing what they eat.

'Golden rice'

Rice is another of the world's major crops. Unfortunately natural rice does not contain much vitamin A. To try and tackle the problem, scientists have developed 'golden rice'.

These rice plants produce vitamin A because scientists have added extra genes to them. A bowl a day of 'golden rice' provides half an adult's recommended daily intake of vitamin A.

If 'golden rice' is successful it could help stop millions of people going blind. However, many scientists worry that there isn't enough evidence to prove that it will really work.

↑ Wheat seeds are used to make flour for bread.

↑ The yellow chemical in 'golden rice' provides vitamin A.

1. Explain what it means for someone to be obese.
2. Why are increasing numbers of people with obesity a serious problem?
3. Give two reasons why the average body mass is increasing in many countries.
4. Suggest two ways of preventing further increases in obesity.
5. Explain why common foods like bread are often fortified.
6. How can we increase the amounts of vitamins in crop plants?

- Most people enjoy eating foods that are sweet, fatty, or salty.
- Consuming too much of these foods can lead to obesity and ill health.
- Many people still suffer from nutritional deficiencies.
- Extra vitamins and minerals can be added to common foods.
- Genes can be added to crop plants to make them produce vitamins.

Review 7.6

1 The picture shows three types of food.

a Which food is mainly carbohydrate? [1]
b Which contains most protein? [1]
c Which provides most energy per gram? Explain why. [2]

2 A food scientist compared the fat content of three different protein sources.

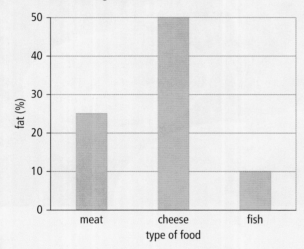

a How many grams of fat does 100 g of cheese contain? [1]
b In the USA, many people want to cut their body mass. Suggest why they are advised to avoid cheese. [1]
c A typical meal contains 100 g of fish or meat. How much more fat does the meat contain? [1]

3 The table below shows three people's muscle mass as a percentage of their body mass.

Person	Muscle (% of body mass)
A	20
B	10
C	30

a A has an above-average amount of muscle. Name the main nutrient needed to build muscle. [1]
b Suggest why B's muscle mass is so low. [1]

4 Crispbread and potato crisps contain different amounts of protein, fat and fibre.

Food	Protein (%)	Fat (%)	Fibre (%)
crispbread	12	8	15
potato crisps	5	28	4

a Predict which food supplies more energy per gram and explain your reasons. [2]
b Explain how the apparatus below could be used to test your prediction. [3]

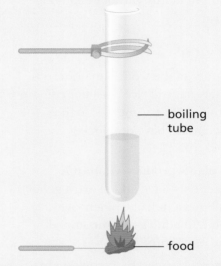

c Suggest two things that must be controlled to make it a fair test. [2]
d Give one reason why the apparatus does not measure all the energy in a piece of food. [1]

5 A medical student tested two different sources of carbohydrate to see how they affected her blood glucose levels. On the first day she ate some sugary sweets for breakfast. The next day she had a bowl of starchy cereal. She measured her blood sugar levels every 30 minutes after eating each food.

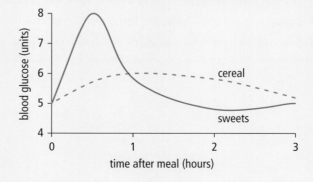

a Which variable did the student test? [1]

b Suggest why she started each experiment first thing in the morning. [1]

c Suggest one other variable she should control, which could also affect blood glucose levels. [1]

d Use the graph to describe the main differences the student found. [2]

e Suggest a reason for these differences. [2]

6 Arjun is an athlete. He needs more energy than most people, and is advised to take in 12 000 kJ per day.

a At least 10% of your daily energy intake should come from protein. Protein releases 16 kJ per gram. How much protein should Arjun take in each day? [2]

b Arjun treats himself to a large chocolate ice cream. It contains 36 g of saturated fat. Fat releases 37 kJ per gram. How many kilojoules does the ice cream give him? [1]

c What percentage of Arjun's recommended energy intake is coming from the saturated fat in this ice cream? [1]

d Explain why Arjun should try not to eat large quantities of saturated fats. [1]

e Arjun rarely eats fish. Explain why oily fish is a good source of fat. [2]

7 Abena made a table to compare the vitamins in some of the fruit and vegetables she eats.

Food	Vitamins (mg/100 g)	
	C	A
orange	45	0.012
banana	9	0.003
spinach	25	0.450
carrot	1.9	0.800

a Scientists estimate that we need 75 mg of vitamin C a day. How many grams of spinach would Abena need to eat to get this much vitamin C? [1].

b The recommended daily intake of vitamin A is 0.9 mg. Suggest two ways Abena could get this much of the vitamin. [2]

c The vitamin C content shown is for raw carrots. How would the vitamin C content change if they were boiled? [1]

8 Rija is a doctor. She sees four patients with nutritional deficiencies. State which nutrient each patient needs more of.

a Patient A finds it difficult to see objects in dim light but her vision is fine in well-lit rooms. [1]

b Patient B feels weak when she exercises and is often tired during the day. [1]

c Patient C is a child whose belly is very swollen. [1]

d Patient D has bleeding gums and is missing a few teeth. [1]

9 Orla has broken her leg twice after falling over. The doctor takes an X-ray of her legs.

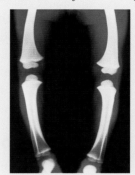

a Describe the appearance of Orla's leg bones. [1]

b Orla spends a lot of time indoors. When she goes outside, she wears a lot of sun block. How does this explain Orla's symptoms? [2]

10 In 1747, James Lind tried to find the best cure for scurvy. He predicted that anything that contained acid would work. He tried each possible cure on two sailors with scurvy. The table shows some of his results.

Source of acid	Condition after 1 week
apple drink	improved
dilute sulfuric acid	worse
vinegar	worse
oranges and lemons	cured

a What variable did Lind change during his investigation? [1]

b What variable did he observe in order to judge the results? [1]

c Does the evidence support the idea that any acid will cure scurvy? [1]

d Other scientists thought that the citric acid in fruits must cure scurvy. They tested their idea, but it did not work. What did the fruits really provide? [1]

8.1 The digestive system

Inside your gut

Objective
- Recognise each part of the alimentary canal and explain its role in digestion

Everything you eat goes into your **gut**, or **alimentary canal**. This long tube winds its way from your mouth to your **rectum**. Food can spend more than a day inside it, but most of your food doesn't get to the end. What happens to it?

The starch, fats, and proteins in your food are made of big complex molecules (see page 92). Your cells can't use them as they are. Inside your gut, these big food molecules are broken down to make small molecules your body can use, such as glucose. These molecules pass into your blood, which carries them to every cell.

⬆ Everything you swallow goes into a long thin tube called your alimentary canal.

Mouth to stomach

Your teeth break up solid lumps of food to make smaller pieces you can swallow. This is **mechanical digestion**. It makes the large molecules in the food easier to break down later.

Your mouth fills with saliva as you chew. Saliva contains an **enzyme**. Enzymes help large molecules break down to form smaller ones. This is **chemical digestion**. The enzyme in saliva helps starch to break down.

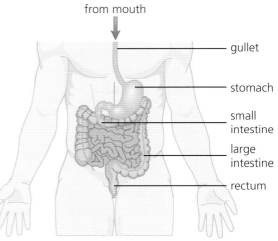
⬆ Your alimentary canal takes food from your mouth to your rectum.

Saliva is also very slippery. It makes food easy to swallow. Then the walls of your gullet squeeze it down to your stomach. This squeezing action of the gut walls is called **peristalsis**.

Inside your stomach

Your stomach blends each meal with acid and another enzyme. The acid destroys microbes and the enzyme begins the digestion of proteins.

Once your meal is a smooth paste, your stomach lets it out. It squirts into the small intestine a bit at a time. Now the pieces of food are smaller, enzymes can mix with the large food molecules more easily.

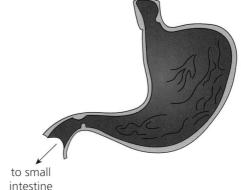
⬆ Your stomach stores food and mixes it with acid and enzymes to make a smooth paste.

In the small intestine

Your **small intestine** is the longest part of your gut. This is where most large molecules get broken down. Its walls squeeze food along and mix it with more enzymes. A lot of these enzymes come from another organ – the **pancreas**.

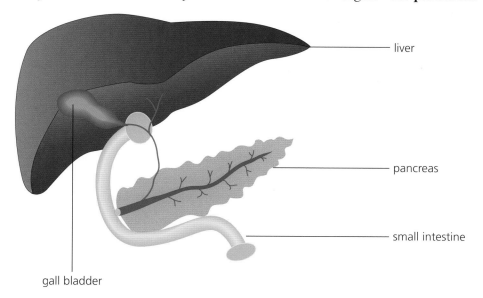

← Your pancreas pours enzymes into the small intestine to aid digestion.

As food leaves the stomach, enzymes from the pancreas pour onto it. Carbohydrate and protein continue to break down and fat digestion starts. The small intestine makes more enzymes to help finish the job.

As food is digested, small molecules are produced. They pass through the small intestine's walls and into your bloodstream. Now you can use them.

Beyond the small intestine

When a meal gets to the **large intestine** there is nothing left but fibre. This part of the gut is full of bacteria. They live off the fibre and make important vitamins that we can absorb.

The walls of the large intestine absorb water into the bloodstream. This turns the mixture of fibre and bacteria into a solid waste called faeces. It is stored in your **rectum** until you are ready to go to the toilet. Then it squeezes out through your **anus**.

Q

1. Explain why most foods need to be digested.
2. Describe the route fibre takes through your body when you eat it.
3. Explain the difference between mechanical digestion and chemical digestion.
4. When your mouth 'waters' it makes saliva. Give two reasons why you need saliva.
5. What happens to food in your stomach?
6. Describe the two main processes that happen in your small intestine.
7. How do small food molecules such as glucose get from your alimentary canal (gut) to your cells?

- Digestion breaks down food into small molecules that can enter your blood.
- Digestion happens in the gut (alimentary canal).
- Teeth crush food into smaller pieces (mechanical digestion).
- Enzymes break down large molecules (chemical digestion).
- Enzymes are produced in your mouth, stomach, pancreas, and small intestine.
- The small intestine absorbs digested food.
- The large intestine absorbs water.
- The rectum stores faeces.

8.2 Enzymes

Catalysts

Life could not exist without enzymes. They control the chemical reactions in your body. Without enzymes, most of these chemical reactions would be too slow to keep you alive.

Chemicals that speed up chemical reactions are **catalysts**, so enzymes are **biological catalysts** – catalysts made by living things. Enzymes can speed up chemical reactions without being changed or used up. The enzymes in your gut make large food molecules break down to form smaller ones.

We can't see the molecules in food but we can use chemicals to detect them. Iodine turns dark blue when it mixes with starch. If warm saliva is added to starch, and left for a few minutes, no blue colour forms with iodine. No starch can be detected because its molecules have broken down.

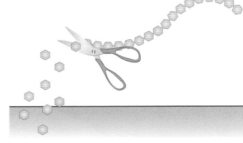

↑ The enzyme in saliva digests starch by helping it to break into glucose molecules.

Objective
- Understand the function of enzymes

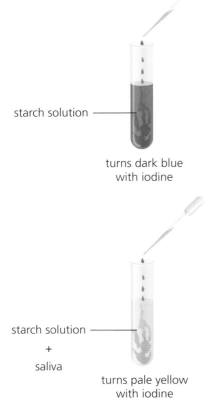

↑ When starch breaks down it no longer turns blue with iodine.

Different enzymes for different jobs

Scissors can cut anything, but enzymes are specialised. Each of the enzymes in your gut can only make one type of molecule break down. The enzyme in saliva is called amylase. It's a type of **carbohydrase** because it breaks down starch – a carbohydrate.

The enzyme produced in the stomach is a **protease** – it makes proteins split into smaller molecules.

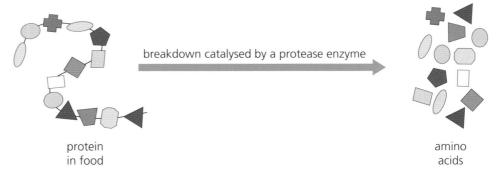

Other enzymes from the pancreas and small intestine complete starch and protein digestion. The pancreas also makes a **lipase** enzyme to make fat molecules break down.

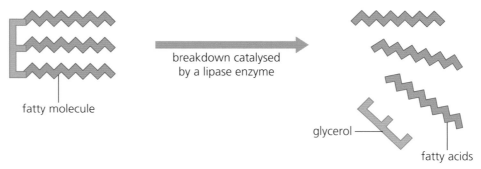

Digestion

Helping our enzymes

In large chunks of food most of the molecules are in the middle. Enzymes can only work on the molecules on the surface of the food. Chewing food breaks it into smaller pieces to give enzymes a **large surface area** to work on. But fat molecules stick together in large globules.

As food enters the small intestine it mixes with **bile** from the **gall bladder** (see page 105). This **emulsifies** fats, which means it breaks them into smaller droplets. The large surface area of the droplets makes it easier for lipase enzymes to break them down.

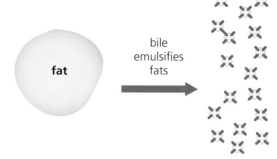

⬆ Lipases work faster with emulsified fats because they have a greater surface area.

Providing the right conditions

Enzymes are giant molecules made from protein. They have to be just the right shape to do their job. Each chemical reaction needs a different enzyme.

If the pH around an enzyme changes, its shape can change. The enzyme becomes **denatured** and stops working. Each enzyme has an **optimum pH** at which the acidity of its surroundings lets it work as fast as it can.

The fluids inside your stomach are very acidic (pH 2) but the inside of your small intestine is slightly alkaline (pH 8). Chemicals from the pancreas, and the chemicals in bile, neutralise food as it leaves your stomach. These differences allow different enzymes to work in each organ.

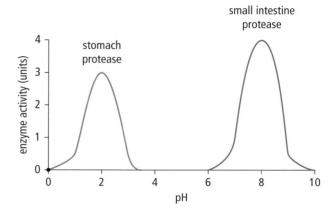

⬆ The proteases that work in your stomach and small intestine work best at different pH values.

High temperatures can also denature enzymes. The enzymes in our bodies work best at 37 °C, but the enzymes in other living things can be very different. Microbes that grow in extreme conditions can have enzymes with optimum temperatures as high as 100 °C or as low as 2 °C.

Q

1. Explain why enzymes are called biological catalysts.
2. Foods need to be broken into smaller pieces before chemical digestion takes place. Why is this?
3. Bile is a green substance which is secreted from your gall bladder. What is its main function?
4. Copy and complete this table:

Nutrient	Enzyme used to digest it	Organ(s) where digestion takes place
carbohydrate		
protein		
fat		

!

- Enzymes are biological catalysts.
- Enzymes in your gut make large molecules break down.
- Each type food needs different enzymes to break it down.
- Starch is digested in your mouth and small intestine.
- Proteins are digested in your stomach and small intestine.
- Fats are only digested in your small intestine.

Extension 8.3

Using enzymes

Objectives
- Understand how enzymes function
- Give some examples of the uses of enzymes

Making cheese

Humans used enzymes for thousands of years before anyone knew what they were. One early use was to turn milk into cheese. To make cheese, most of the protein in milk is turned into a solid curd. This can be done by adding plant juices, or an animal's stomach juices, to the milk. The curds are then pressed and left to mature to produce solid cheese.

↑ Protease enzymes convert milk to solid curds during cheese-making.

When scientists tested the plant and animal juices they found protease enzymes in them. Since the 1980s, cheese-makers have been able to buy a pure protease enzyme made by bacteria. This has allowed much more cheese to be made.

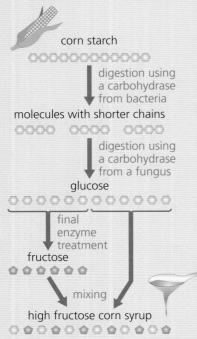
↑ Three enzymes convert starch to high-fructose corn syrup.

Making sweeter sugar

We used to get our sugar from fruit, honey, sugar cane, or sugar beet. Now most of the sugar in processed food and drinks is made from corn starch, which is cheaper. Carbohydrase enzymes from bacteria and fungi are used to break the starch down and make glucose. Then a third enzyme turns half the glucose into fructose. Fructose is much sweeter than other sugars so less needs to be used to make food taste good.

Improving foods with lipases

Lipases make many foods taste better – especially cheeses. When lipases break down fats they release fatty acids. Then more chemical reactions turn these fatty acids into substances that make foods taste better.

All fat molecules contain glycerol bound to three fatty acids. The fatty acids determine the properties of the fat molecules and how much they are worth in the food industry. Lipases can improve fats by changing some of the fatty acids in them. For example, cheap vegetable oils can be made into hard butter or soft margarine rich in essential omega-3 and omega-6 fatty acids.

The table shows the effects of lipase treatment on different foods.

Food	Effect of lipase treatment
bread	lasts longer/tastes better
dairy foods	improves taste
drinks	improved aroma (smell)
sauces	smoother texture
meat/fish	fat removal
health foods	adds essential fatty acids

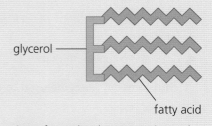
↑ A fat molecule's properties and value depend on the types of fatty acids it contains.

Digestion

How enzymes work

Enzymes speed up chemical reactions by making them happen more easily.

The sugar we get from sugar cane is sucrose. Its molecules contain glucose and fructose. If you eat sucrose, enzymes along the walls of your small intestine break it down and release the two smaller sugars.

The sucrose binds to the part of the enzyme called its **active site**. This is shaped to fit the sucrose and hold it in place. Every enzyme has a different shape and most enzymes will only speed up one sort of reaction.

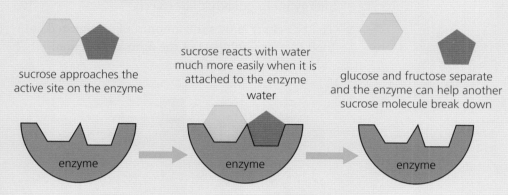

← This enzyme breaks down sucrose by speeding up the reaction between sucrose and water.

Sucrose needs to react with a water molecule to split into glucose and fructose. The enzyme makes this reaction much easier. As soon as the reaction has finished, it lets the separate glucose and fructose go. The enzyme is left unchanged and ready to bind to another sucrose molecule.

Enzyme-catalysed reactions are fast. Each enzyme molecule breaks down thousands of sucrose molecules per second. A small amount of enzyme can make huge quantities of a chemical react.

All the enzymes in your digestive system break large molecules down by helping them to react with water. Most of the enzymes used in industry do too. But enzymes can also speed up other types of reaction. They can help build large molecules as well as break them down.

Why use enzymes?

Microbes make a huge variety of enzymes. They make impossible chemical reactions possible and expensive reactions cheap. They are *nature's gifts to us*. More than 3000 different enzymes have been identified and that number is growing all the time. The right enzyme can help us to make anything we need.

---

1 Copy and complete this table:

Type of enzyme	Use in food manufacture
carbohydrase	
protease	
lipase	

2 Maltose is a sugar. It contains two glucose molecules joined together. Draw diagrams to show how an enzyme could break it down.

3 Explain why enzymes are only needed in small amounts.

4 Scientists all over the world are investigating the enzymes different microbes produce. Explain why this work is important.

- Enzymes are used to make cheese, produce sugar, and improve foods.
- Every enzyme has an active site that binds the molecules it works with.
- Enzymes don't change during reactions. They can be used over and over again.

Review 8.4

1 Write the correct letter from the diagram for each of the following parts of the alimentary canal.

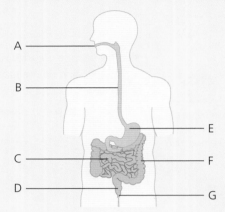

   a stomach [1]
   b small intestine [1]
   c gullet [1]
   d mouth [1]
   e large intestine [1]

2 Name the part of the alimentary canal that fits each of the following descriptions.
   a contains strong acid to help kill the micro-organisms in food [1]
   b its muscular walls use peristalsis to help deliver food to the stomach [1]
   c absorbs a lot of water to produce solid waste from the food we eat [1]
   d mixes food with enzymes that break down protein [1]
   e grinds lumps of food up and mixes it with saliva [1]
   f where nutrients are absorbed into the bloodstream [1]

3 Digestion begins at the mouth.

   a Explain how teeth aid digestion. [2]
   b Give two ways in which saliva aids digestion. [1]

4 Many sports drinks contain glucose. Athletes can also get energy from starch in their diets.

part of a molecule of starch

molecules of glucose

Explain why glucose molecules can provide energy more quickly than starch. [2]

5 The bar chart shows the amount of undigested food leaving each part of the alimentary canal.

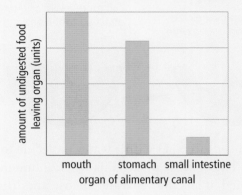

   a Where does most digestion take place? [1]
   b What happens to food that cannot be digested? [3]

6 The flow chart shows how food passes through the alimentary canal.

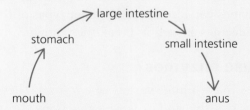

   a Which two parts are in the wrong order? [1]
   b Which three parts produce enzymes? [1]

7 The graph shows how the amount of digested food in the blood leaving the small intestine changed after a student ate a meal.

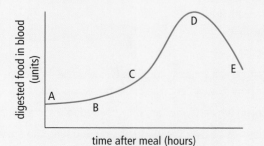

Choose the correct part of the graph for each of the following.

a Food begins to enter the small intestine. [1]

b Food begins to enter the large intestine. [1]

8 A protein solution was added to two test tubes. Then water was added to one test tube and a protein-digesting enzyme was added to the other. After 3 hours a special indicator was added to both tubes. It turns lilac when protein is present. The table shows the results.

Test tube	Indicator colour
protein + water	lilac
protein + enzyme	blue

a State what was observed or measured in this investigation. [1]

b Suggest an explanation for the results. [2]

9 A mixture of starch and glucose solution was placed in a model gut in a beaker of water.

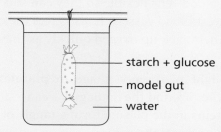

After 30 minutes the water in the beaker was tested for starch and glucose. Only glucose was present.

a Suggest how glucose got out of the model gut. [1]

b Suggest why no starch was detected in the water. [2]

In a second experiment the model gut was filled with starch and saliva.

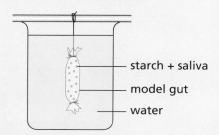

After 30 minutes the water in the beaker was tested for starch and glucose. Glucose was present but not starch.

c Explain what happened to the starch. [2]

d In the model, what does the water in the beaker represent? [1]

10 Cubes of protein were dropped into two test tubes. One tube contained protease from human stomach juices. The other contained boiled protease. The tubes were both kept at 37 °C for 4 hours.

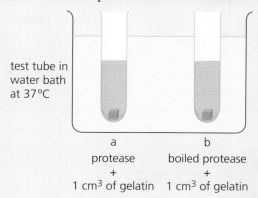

a Suggest why the tubes were kept at 37 °C. [1]

b State what was changed and what was observed. [1]

c List two variables that were controlled. [2]

d Describe and explain the results in the table below. [2]

Enzyme added	Appearance of protein
normal protease	all digested
boiled protease	none digested

In a second experiment normal protease was added to a cube of protein and an identical cube that had been cut into small pieces.

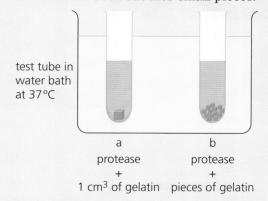

The tubes were checked every 10 minutes.

e Describe and explain the results in the table below. [2]

Protein	Time taken to digest protein (hours)
1 cube	3.5
many pieces	1.5

f Which part of the digestive system breaks food into smaller pieces? [1]

9.1 Blood

Blood cells

Objectives
- List the components of blood
- Describe the function of each component

Under a microscope you can see that blood isn't completely red. It is a watery liquid called **plasma** full of tiny cells. The largest cell in this microscope image is a **white blood cell**. It is stained purple to make it stand out. Cells like this help to destroy the micro-organisms that cause infectious diseases.

The other cells are **red blood cells**. There are about 5000 million in each cubic centimetre of your blood. The SEM image on the left shows their 3D shape more clearly.

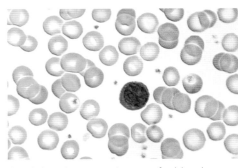

↑ Light microscope image of a blood smear (magnified 1150 times).

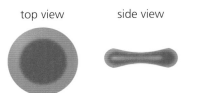

↑ The red blood cell's biconcave shape makes it very flexible and provides a large surface area to take in oxygen.

A light microscope makes the red cells look flat and hollow. In reality they are biconcave, which means both their sides curve inwards.

This shape is useful. It makes red blood cells flexible so they can squeeze through **capillaries** – your narrowest blood vessels. It also gives the cells a **large surface area**, which helps them pick up oxygen quickly as they pass through your lungs.

The smaller purple objects in the light microscope image are **platelets**. If you cut a blood vessel the platelets gather around the damaged area to stop blood leaking out. At the same time proteins in the plasma form a tangle of fine strands. These trap red blood cells and seal the wound.

Plasma

Your blood transports lots of substances, and plasma carries them all except oxygen. Plasma transports digested food from the small intestine, carbon dioxide from respiring cells, and **urea**.

Urea is a waste product made in your liver. Your blood plasma carries urea to your **kidneys** to be excreted in your **urine**.

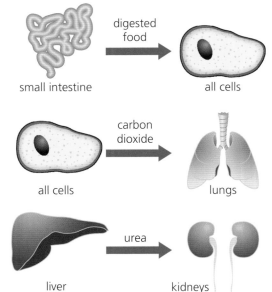

↑ Blood plasma transports substances around your body.

Capillaries

Every cell in your body is close to a capillary so it can collect oxygen from passing red blood cells. Dissolved substances always move from where they are concentrated to where their molecules are more spread out. This is **diffusion**. So oxygen and glucose **diffuse** into cells and carbon dioxide diffuses into blood.

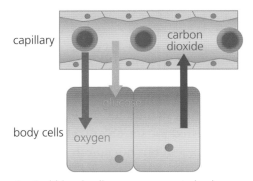

↑ Red blood cells carry oxygen to body cells. Plasma brings glucose and takes away carbon dioxide.

Circulation

Blood tests

If you are ill, your doctor may ask you to have a blood test. Many diseases can be detected by looking at blood under the microscope.

Some people are born with sickle-cell anaemia. They have faulty haemoglobin. It makes some of their red blood cells curve into long thin C-shapes. These curved red cells get stuck in narrow blood vessels. They may block them completely and cause a lot of tissue damage and pain.

The protozoa that cause malaria (see page 151) spend part of their lives in red blood cells. They feed on the haemoglobin, reproduce, and then burst out to invade a new red blood cell. Infected red blood cells are sticky and can block small blood vessels.

The protozoa that cause sleeping sickness spend part of their lives in human blood. They cause fever, headaches, joint pains, and itching. Later they invade their victim's brain and cause the tiredness and confusion that gives the disease its name.

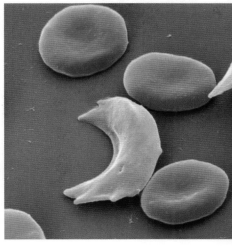
↑ This blood is from a child with sickle cell anaemia

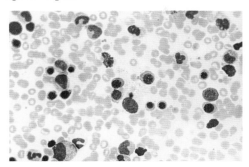
↑ The protozoa invading these red cells cause malaria

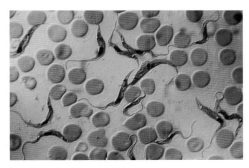
↑ The protozoa in this blood cause sleeping sickness

1 Copy and complete this table to show the function of each component of blood.

Blood component	Main function
red blood cell	
white blood cell	
platelet	
plasma	

2 Draw a side view of a red blood cell to show its biconcave shape.

3 Give three reasons why red blood cells are good at carrying oxygen.

4 Copy and complete this table to show what blood transports.

Substance	From	To
oxygen		
carbon dioxide		
digested food		
urea		

5 Suggest why micro-organisms can grow and divide very quickly if they get into blood.

- Blood contains liquid plasma, platelets, and red and white blood cells.
- Digested food, carbon dioxide, and urea are carried in the plasma.
- Oxygen is carried in red blood cells bound to haemoglobin.

Extension 9.2

Anaemia

Objectives
- Make simple calculations
- Recognise what we can learn from blood tests

Tests

Sara never seems to have any energy. Her doctor thinks she might be **anaemic**. He has sent her for a blood test. If you're anaemic your blood cannot carry enough oxygen. This makes you feel very tired all the time. The nurse at the hospital is taking a blood sample to send to the laboratory.

A haematologist (see page 23) puts Sara's blood in a machine called an electronic blood-cell counter. The counter sucks up a small amount of her blood. Each cell in the blood is counted and measured as it moves through the machine. You can count blood cells in a drop of blood by viewing it under a microscope, but it is quicker to use a machine.

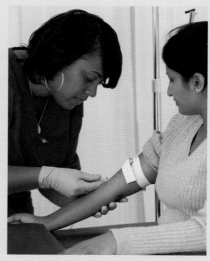

↑ Doctors often request blood tests to help confirm their diagnosis.

Results

The counter works out the number of cells in each cubic decimetre (dm^3) of Sara's blood. As with all body measurements, people show a lot of variation. However, most people's red cell counts lie within a specific range. The range for men is higher than the range for women.

Sara has 4 dm^3 of blood altogether and her red cell count is 3 billion cells per dm^3.

Normal red cells counts (billions per dm^3)	
Men	Women
4.5–5.6	3.8–5.0

The machine also finds out how much haemoglobin there is in each dm^3 of blood. Sara's blood has 93 grams of haemoglobin per dm^3.

Normal haemoglobin levels (grams per dm^3)	
Men	Women
130–170	110–150

Observations

The haematologist checks the appearance of Sara's blood. First she spreads a single drop out very thinly to make a blood smear. Then she compares Sara's red cells with cells from a normal blood sample.

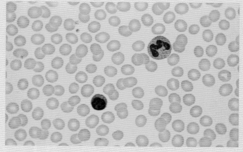

↑ Normal blood – magnified 1000 times

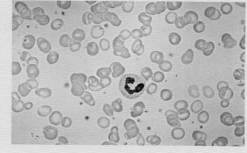

↑ Sara's blood – magnified 1000 times

Circulation

More checks

The haematologist also measures how much of Sara's blood is made up of red cells. The red cells are separated out so that they pack together at the bottom of a tube. Her plasma forms a clear layer on top and there is a fuzzy layer of white blood cells in the middle.

A low packed cell volume would mean that Sara has fewer red cells than normal, or that her red cells are smaller than normal.

Normal packed cell volumes (%)	
Men	Women
38–50	36–46

Diagnosis

Sara's test results are sent to her doctor. They confirm that she is anaemic. She has fewer red cells than normal and less haemoglobin. She feels tired all the time because her red cells can't carry enough oxygen.

The doctor explains what she needs to do. 'Your body needs iron to make haemoglobin. It looks as if you don't have enough iron in your diet. Try eating more eggs, red meat, and dairy products, and take these iron tablets twice a day. If you don't start feeling better, come back. We can do more tests.'

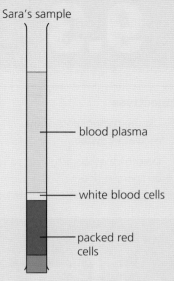

↑ Blood can be separated into cells and plasma.

↑ Doctors use blood tests results to decide the best treatment for their patients.

1 Do all healthy people have the same red cell count?
2 Is Sara's red cell count normal?
3 Does Sara have a normal amount of haemoglobin?
4 What do you notice about the colour of Sara's red cells?
5 Follow these three steps to compare Sara's blood cells with a normal sample:
   a Measure the diameter (distance across the middle) of 5 cells from Sara's blood.
   b Add the measurements, then divide the total by 5 to get an average value.
   c Now do the same with the cells from the normal blood sample.
6 Describe any difference you find between Sara's cells and the normal ones.
7 Follow these three steps to find out what percentage of Sara's blood is made of cells:
   a Measure the height of the packed red cells in Sara's sample.
   b Measure the total height of the cells plus plasma.
   c Use this equation to calculate the result:

   $$\text{Packed cell volume \%} = \frac{\text{height of red cells} \times 100\%}{\text{total height}}$$

8 Does Sara's packed cell volume agree with the results from her other tests?

- Hospitals use blood tests to find causes for ill health.
- Anaemia reduces the amount of haemoglobin in blood, so it carries less oxygen.

9.3 The circulatory system

Running

Martin wins marathons. He can keep on running for hours. The key to his success is an efficient circulatory system. His blood picks up oxygen quickly and delivers it to his muscles.

Using a model

Blood makes tissues look red in real life, but scientific **models** use colour to make the circulatory system clearer. The blood in the left side of Martin's heart (on the right in the image) is red to show it is full of oxygen. This side pumps hard, so blood shoots through the red arteries to every corner of his body. You can see what a heart really looks like on page 20.

Objectives
- List the components of the circulatory system
- Describe the job each component does

↑ A marathon runner needs an efficient circulatory system.

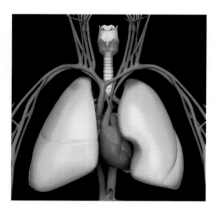

↑ The right side of your heart (blue) pumps oxygen-poor blood through your lungs. The left side (red) pumps oxygen-rich blood around the rest of your body.

Arteries, capillaries, and veins

Arteries are like high-speed highways for red blood cells. They take blood from the heart to capillaries in your muscles, bones, and organs. Capillaries are like roads, and your cells are the houses. Just as roads take cars to houses, capillaries channel blood to cells. You have about 100 000 km of blood vessels, and most of them are capillaries.

↑ Arteries branch to make the networks of capillaries that run through all your tissues.

As capillaries leave each organ, they join to make veins. The veins in the model are coloured blue to show they are short of oxygen. They bring blood back to your heart. Then the right side of your heart (left on the picture) pushes it through your lungs to pick up oxygen.

Circuits

So blood travels from your heart to your lungs, and back to the other side of your heart. Then another push sends the oxygen-rich blood through capillaries in the rest of the body. A wall of muscle down the middle of your heart stops blood taking a short-cut.

On each trip, some blood flows through your small intestine to top up its glucose levels.

↓ On each trip round the body, some blood visits the small intestine take in more glucose.

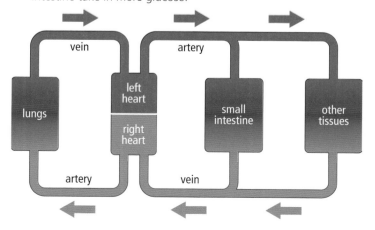

Under pressure

If you fall and cut you knee, you usually damage a capillary. The bleeding soon stops. If blood flows out in a smooth stream, you have cut a vein. If you cut an artery, blood spurts out under very high pressure and the bleeding is very hard to stop.

Each heartbeat gives blood a hard push, so the blood in arteries is under pressure and moving fast. Arteries have thick muscular walls to withstand the pressure of the blood leaving the heart.

Blood loses pressure as it pushes through your capillaries. It drifts back to the heart, through wide veins, under very low pressure. Valves in veins keep it flowing in the right direction. They slam shut if blood tries to slip backwards as it climbs up your arms and legs.

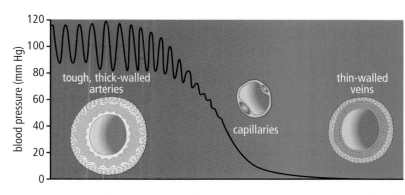

⬆ Arteries need thick muscular walls to withstand high pressures as blood leaves your heart

Meeting demands

Your heart can make blood circulate five times faster when you run. But that's still not enough to supply all the glucose and oxygen your muscles need. So your circulation changes to match each tissue's need.

The chart shows how the blood sent to each organ changed when an athlete started to run.

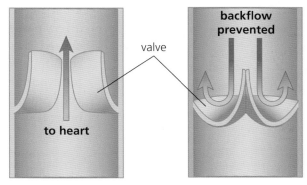

⬆ Valves keep the blood in veins flowing towards the heart.

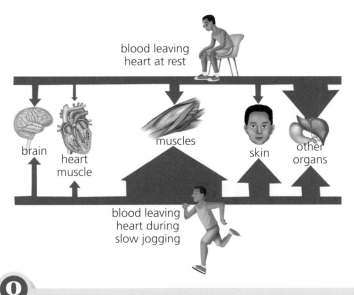

⬅ When you run more blood is sent to your muscles and less to organs like your digestive system.

Q
1. On each trip around the body, blood passes through your heart twice. Explain why.
2. Which side of the heart sends blood through your lungs to collect oxygen?
3. Some blood has just left your muscles. Its oxygen content is low. Describe the route it takes to get back to your muscles.
4. Copy the diagram of your circulatory system from page 116. Label the vein where blood has most oxygen, and the artery where it has least oxygen.

- Blood leaves your heart in arteries and returns in veins.
- The right side of your heart pumps blood through your lungs. The left side sends blood to every other part of your body.
- Capillaries bring blood close to every cell.

Enquiry 9.4 Identifying trends

Pulse rate and exercise

Ahmed is playing football. The more he runs, the more oxygen his muscles need. His heart beats harder and faster to send more blood to his muscles. At the end of the match he times the **pulse** in his wrist. He counts 170 beats per minute. That's high.

You can feel a pulse wherever an artery passes over a bone. It is caused by the pressure increase each time your heart pumps blood into your arteries. So measuring your pulse shows how fast your heart is working.

Objectives

- Understand how the circulatory system responds to exercise
- Identify trends and patterns in results

↑ A footballer's heart beats harder and faster during a game.

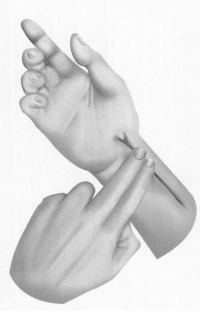

↑ Pressure from each heart beat causes a pulse wherever an artery passes over bone.

Spotting a trend

Ahmed checks his pulse for another 8 minutes. He can see that his heart rate goes down as the time after exercise increases.

Time after exercise (minutes)	Heart rate (beats per minute)
0	170
2	124
4	94
6	80
8	75
10	74

Ahmed plots the results on a graph and draws a **curve of best fit**. Now he can see more detail. His heart rate goes down quickly at first. Then it drops more slowly.

Ahmed's coach sees the graph. He says it would look better if Ahmed was fitter. He suggests a training program.

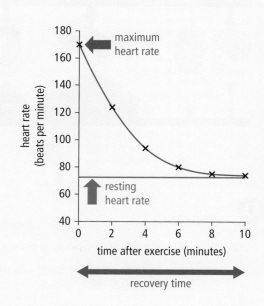

Comparing graphs

After 8 weeks of training Ahmed feels fitter – but will his new results please his coach? He adds them to his graph. They look different. His coach is pleased. He points out three differences between Ahmed's two curves. These show he is much fitter than he was 8 weeks ago.

Maintaining supplies to muscles

Everyone's heart rate increases when they exercise. Ahmed's training made his heart more efficient. His heart muscle is stronger now. It pumps more blood out each time it beats, so it doesn't need to pump so fast. He also recovers much faster when he stops exercising.

Another variable

Scientists can measure how much oxygen blood carries.

Ahmed's blood is normal. Each dm^3 carries $200 \, cm^3$ of oxygen.

Toey has anaemia so her blood can't carry so much oxygen.

Aram lives in the mountains. His body has adapted to the thin mountain air by making more red blood cells. When he comes down from the mountain his blood can carry more oxygen.

Ahmed, Toey, and Aram cycle at the same speed for 5 minutes and then measure their heart rate. The graph shows the relationship between the volume of oxygen their blood can carry and their maximum heart rate.

Their results all lie close to a **line of best fit**. This means that there is a **correlation** between the two variables. It is a **negative correlation** because the maximum heart rate goes down as the oxygen content of blood increases.

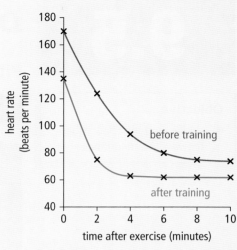

↑ Ahmed is much fitter after training.

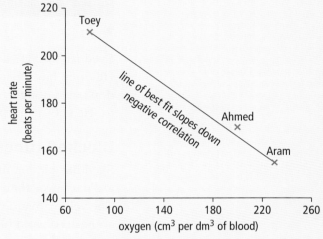

↑ There is a negative correlation between the heart rate and the volume of oxygen in the blood. When one increases the other decreases.

Q

1 Copy the shape of Ahmed's first graph and describe the pattern it shows.

2 Copy and complete this table to summarise Ahmed's results.

	Before training	After training
resting heart rate (beats/minute)		
maximum heart rate (beats/minute)		
recovery time (minutes)		

3 Describe the correlation between the volume of oxygen blood can carry and heart rate during exercise.

!

- Your heart beats harder and faster during exercise.
- Regular exercise makes your heart more efficient.
- This decreases your heart rate and shortens your recovery time.
- Graphs can show relationships between variables.
- There is a correlation between two variables if all the points on a graph lie close to a line of best fit.

9.5 Diet and fitness

Fitness

Objective
- Understand the relationship between diet and fitness

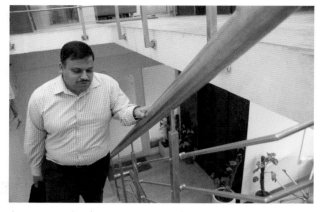

↑ A poor circulatory system can leave you short of energy.

Syed isn't fit. He doesn't have as much energy as he did when he was young. He often has to stop and rest. Less blood is able to reach his muscles so they don't get enough glucose and oxygen. They cannot respire as fast as they need to. He gets a sharp pain in his leg muscles when he tries to run.

Blocked tubes

Syed has plaque in his arteries. They are narrower than they should be. The yellow plaque is a mixture of fat, cholesterol, and blood cells. Plaque can start forming when you are in your teens and it gradually builds up as you age. The plaque can reduce the blood flow to your muscles. If plaque builds up in the arteries supplying your heart, it may not get enough glucose and oxygen.

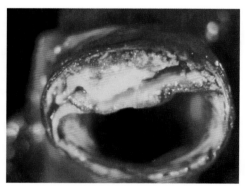

↑ Plaque narrows arteries and raises your blood pressure.

Syed's doctor measures his blood pressure to monitor his condition. The pressure keeps changing. When Syed was younger his blood pressure was 120/80. The high number is the pressure in his arteries when his heart pumps his blood out. The low number is the pressure between beats. Now Syed's blood pressure is 160/95. His heart has to pump much harder to push blood through his narrow arteries.

High blood pressure is serious. It can damage your heart, brain, vision, and kidneys.

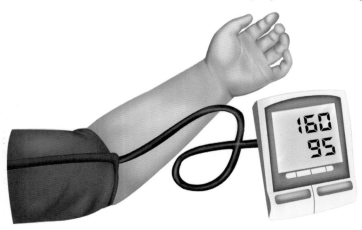

↑ Blood pressure naturally rises with age, but a very high blood pressure is dangerous.

Circulation

Heart attacks and strokes

Plaque can break away from an artery wall and travel in the blood to block an artery somewhere else. Plaque can also tear. When this happens platelets stick to it and form a blood clot, which can also block an artery.

If the blocked artery supplies your heart muscles, a blockage causes a heart attack. A heart attack victim usually has chest pains and pains in their left arm. They feel as if their body is being squeezed. They also feel sick and struggle to breathe.

If the blood supply to your brain is blocked, it causes a stroke. Many stroke victims die. Some are left paralysed. Prompt medical treatment can reduce the brain damage a stroke causes. Most strokes cause these symptoms straight away: the victim's face droops to one side; they can't hold their hands in the air; and their speech is harder to understand.

Fortunately, a healthy lifestyle lowers the chance of having a heart attack or stroke.

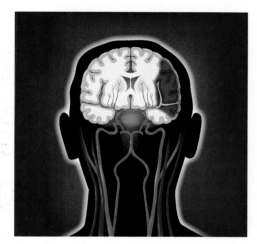

⬆ When a blood clot blocks an artery in the brain, large numbers of nerve cells die. The scan shows this on the right-hand side of the image.

Staying healthy

Some people have a greater risk of having a heart attack or stroke than others. But why? Scientists all over the world compared thousands of people for many years to look for clues. Those who died early had more of the risk factors in the table. More young people are having heart attacks than ever before.

Risk factors
smoking
being overweight
taking no exercise
unhealthy food
high salt intake
excess alcohol
inherited genes

Eating right

Syed doesn't smoke or drink. His doctor advises him to improve his diet and start doing some exercise. The first thing he plans to do is cut down his salt intake.

Regular exercise and a good diet stop you gaining weight. They also lower your risk of developing high blood pressure. To make this simpler doctors have made a list of foods to eat and foods to avoid.

Nutrients	Try to eat	Aim to avoid
fat	oils, nuts, seeds, fish	processed foods, saturated fats in meat
carbohydrates	salads, beans, fruits	sweets, sugary drinks, potatoes, white bread
protein	fish, chicken	red meat, sausages

Q

1. Draw diagrams to show: a) a normal artery; b) an artery with a lot of plaque.
2. What is the plaque made from?
3. Why is it a problem if you have plaque in your arteries?
4. Explain how blood pressure measurements can show whether someone's arteries are getting narrower.
5. What is the same about heart attacks and strokes?
6. Which risk factors can people avoid by changing their diet?

- Plaque in arteries puts a strain on your heart and raises your blood pressure.
- Lack of exercise, a poor diet, smoking, and drinking alcohol increase your risk of having a heart attack or stroke.

Review 9.6

1 The diagram shows three of the components of blood.

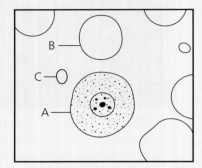

Give the letter of the component that has each of these functions:

a aids blood clot formation [1]

b helps to destroy micro-organisms [1]

c carries oxygen. [1]

2 The arrows show substances entering and leaving a capillary in a muscle.

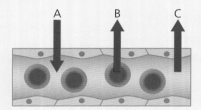

Give the letter of the arrow that shows the direction in which each of the following move:

a glucose [1]

b carbon dioxide. [1]

3 Three systems work together to deliver essential supplies to cells.

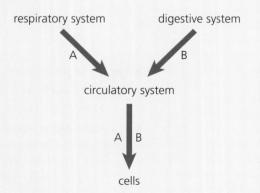

Name substances A and B. [2]

4 Blood cells are suspended in a watery liquid.

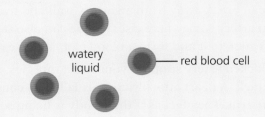

a Name the watery liquid. [1]

b Name two substances carried in the watery liquid. [2]

5 The diagram shows how blood circulates around the body.

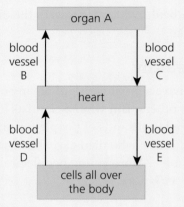

a Name organ A. [1]

b Give the letters of the two arrows that represent veins. [2]

6 A student's heart rate was measured once a minute for 30 minutes. Then the results were plotted on a graph.

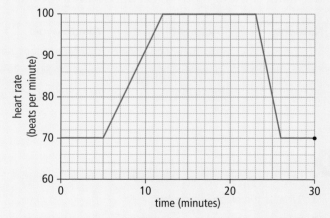

a According to the graph, when did the student start to do some exercise? [1]

b Explain why her heart rate changes during exercise. [2]

c When did the student stop exercising? [1]

Circulation

7 The diagram shows some capillaries.

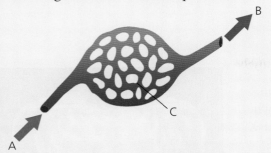

a Where has the blood in A come from? [1]

b Which blood vessel contains most carbon dioxide, A, B, or C? [1]

c Which blood vessel has the thinnest walls? Explain why. [2]

8 A sports scientist measured two students' heart rates when they were resting and after they ran for 2 minutes at the same speed.

Student	Heart rate (beats per minute)	
	Sitting	Running
A	76	120
B	62	80

a Explain why their heart rates were higher when they were running. [1]

b Which student is fitter? [1]

c How could a student improve the efficiency of their circulatory system? [1]

9 The tables show the amounts of oxygen and carbon dioxide in blood from sides A and B of the heart.

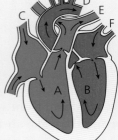

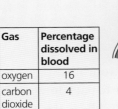

Gas	Percentage dissolved in blood
oxygen	16
carbon dioxide	4

Gas	Percentage dissolved in blood
oxygen	21
carbon dioxide	0.04

a Which side of the heart pumps blood to the lungs, A or B? [1]

b Which two letters show arteries? [2]

10 The diagram shows how blood circulates.

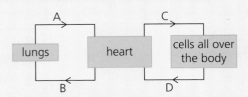

The bar chart shows the amount of oxygen in blood at positions A–D, but they are in the wrong order.

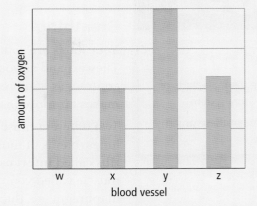

a Which artery did blood sample X come from? Answer A, B, C, or D. [1]

b Which vein did blood sample Y come from? Answer A, B, C, or D. [1]

11 The diagram shows blood entering and leaving the heart.

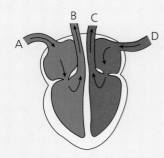

a The blood in vein D has the highest oxygen concentration. Explain why [1]

b Which artery carries blood with a low oxygen concentration? [1]

c Which artery takes blood to your arms and legs? [1]

12 This diagram shows part of a vein.

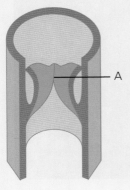

a Name structure A. [1]

b Describe what structure A does. [1]

10.1 Lungs

Breathing

Objectives
- Recognise the main parts of the respiratory system
- Describe what these organ systems do

Your lungs are soft and spongy. They expand as you breathe in and contract as you breathe out. But what makes air move in and out like this?

Two different muscles control your lungs: bands of muscle between your ribs called **intercostal** muscles and a muscle called your **diaphragm**. Your diaphragm is a thick sheet of muscle. It separates your heart and lungs from the rest of your organs.

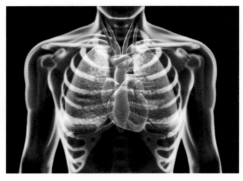

↑ Ribs protect your soft, spongy lungs.

When these muscles contract, your chest gets bigger. Your ribs move up and out and your diaphragm flattens. That makes air rush in through your mouth and nose. When the muscles relax, your chest gets smaller and air flows out again.

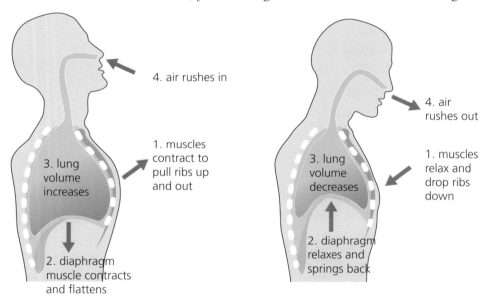

Inside your lungs

A strong tube called the **trachea** carries air down from your nose and mouth. Then it divides into two **bronchi** to take air to each lung. On the model in the photograph, the clear tube in the middle is the end of the trachea.

Each **bronchus** keeps dividing like the branches of a tree. It forms smaller and smaller tubes. Each tiny tube is a **bronchiole**. The model only shows the largest ones. There are millions of them. At the end of each bronchiole is a cluster of tiny air spaces called **alveoli**. They are too small to show on the model.

Threaded between the bronchioles is another set of branching tubes, shown red on the model. These are arteries, carrying blood to the alveoli.

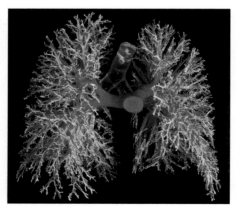

↑ Inside your lungs your airways (white) and blood vessels (red) spread out like the branches of a tree.

Alveoli

Each alveolus is surrounded by a network of capillaries. The blood arriving at the capillaries in the lungs is short of oxygen. It is coloured blue in the diagram to show this lack of oxygen. In reality, it is just a slightly darker red than blood that is rich in oxygen.

The walls of the capillaries and alveoli are very thin. Oxygen moves from the air inside the alveoli into the blood in your capillaries. It moves easily by diffusion. At the same time, carbon dioxide moves in the opposite direction – from your blood plasma to the air. This swopping of oxygen and carbon dioxide is **gas exchange**.

Blood spends less than a second in the capillaries around your alveoli. As your red cells race through, they pick up as much oxygen as they can carry. They are normally full of oxygen when they leave the lungs to return to the heart. The oxygen binds to haemoglobin – the red protein inside red blood cells.

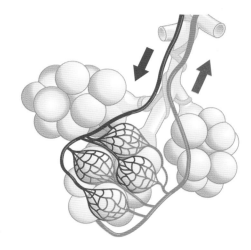

Speeding up diffusion

Alveoli are designed to make diffusion fast. The light microscope image shows how thin their walls are. These thin walls bring air and blood as close together as possible. Red blood cells squeeze along the capillaries around them. The inner surface of each alveolus is covered with a thin layer of moisture. Oxygen dissolves in this and then diffuses through the thin walls of the alveolus and capillary.

Lungs are good at taking in oxygen because they provide a large surface area. You have at least 300 million alveoli. If you spread out their walls they could cover half a tennis court. At any moment, a quarter of a mug of blood is flowing over this very large surface area. It is spread extremely thinly, so oxygen can diffuse into it quickly.

⬆ Red blood cells squeeze through narrow capillaries around the thin walls of the alveoli – magnified 100 times.

Q

1. Name the two different sets of muscles that make you breathe.
2. Explain how they work together to make air enter your lungs.
3. Most organs sink in water, but lungs float. Suggest why.
4. Imagine you are an oxygen molecule that has just been breathed in. Describe the route you take to get into the blood.
5. Blood that is short of oxygen is often coloured blue on diagrams. Which blood vessels should be coloured blue, those that carry blood to the lungs or those that carry it back to the heart?

- Your diaphragm and muscles between your ribs make air move in and out of your lungs.
- It travels through the trachea, bronchi, and bronchioles to the alveoli.
- Gas exchange between air and blood takes place in the alveoli.
- Oxygen passes into the blood and carbon dioxide leaves.

10.2 Respiration and gas exchange

Objectives
- Describe aerobic respiration
- Distinguish between respiration and gas exchange

Holding your breath

In 2012 Tom Sietas set a new world record for holding his breath under water. He lasted more than 22 minutes. Don't try this at home! The average person can only hold their breath for about 18 seconds.

To achieve his record-breaking time, Tom filled his lungs with pure oxygen first. Once he was in the water he kept totally still. The cold water made his heart beat more slowly and cut the blood supply to his arms and legs. That left enough oxygen in his blood to keep his heart and brain cells alive.

↑ Tom Sietas can hold his breath for longer than anyone else.

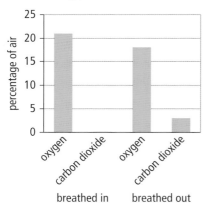

⬇ The air you breathe out still has some oxygen in it.

Why did Tom use pure oxygen?

Only 21% of the air is oxygen. The rest is mainly nitrogen. There is also a small amount of carbon dioxide and other gases. So a lung full of pure oxygen supplies more oxygen than a lung full of air.

Gas exchange in the lungs changes the composition of the air. Breathed-out air has less oxygen and more carbon dioxide than breathed-in air. Breathing renews the air in the lungs. It keeps the oxygen concentration in the alveoli as high as possible to make diffusion into the blood as fast as possible.

Why did Tom keep still?

Cells all over your body use oxygen to release energy from food molecules such as glucose. The chemical reaction they use is called **aerobic respiration**.

↑ Aerobic respiration releases energy from glucose and oxygen.

Cells use the energy from respiration to keep themselves alive, stay warm, grow, reproduce, and move. Muscle cells use a lot of the energy from respiration for movement. By keeping very still, Tom reduced his use of energy, cut respiration to a minimum, and saved oxygen.

Getting more oxygen to your cells

When you sit still like Tom, you need less oxygen. When you start to exercise your body's demand for energy goes up, so respiration has to speed up very quickly. Your cells need more oxygen so gas exchange has to increase.

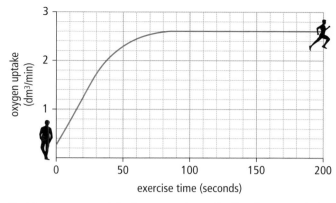

↑ Your oxygen uptake increases dramatically when you start to run.

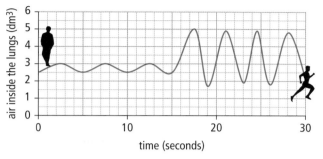

↑ Your lungs breathe faster and deeper when you start to run.

When you are sitting down your lungs will be taking in about 0.5 dm³ per breath. A special graph called a spirogram shows how the volume of air in the lungs changes as you breathe in and out. When you exercise, your lungs increase gas exchange by breathing faster and deeper. This provides the extra oxygen that your cells need.

Coping at high altitudes

The higher you go, the less oxygen there is in a lungful of air. That makes climbing high mountains very difficult. When there is less oxygen in the alveoli, red blood cells can't fill up with oxygen properly so they carry less oxygen around the climber's body.

Given time, most people can adapt to low oxygen levels. Their bodies make more red blood cells and grow extra capillaries in their muscles. Athletes often train at high altitudes. The extra red blood cells let them use oxygen faster, so they can respire faster and run faster – but the effect wears off after a week or two.

Most climbers on Mount Everest employ local people to carry their luggage. These people's ancestors lived high up. Those with helpful genes were more likely to survive and pass these genes to their children. So most local people cope well with low oxygen levels.

↑ People whose ancestors lived at high altitudes are adapted to the lack of oxygen there

Q

1. What is the name of the reaction that releases energy inside cells? Write the word equation for this reaction.
2. Inside your lungs air gets warmer and wetter. How else does it change?
3. Your body needs more oxygen when you are exercising than when you are not exercising. Why is this?
4. Use the graphs on this page to answer these questions.
   a. What was the runner's oxygen uptake when he was resting?
   b. How much higher was his oxygen uptake after he had exercised for 100 seconds?
   c. How many breaths did he take in 15 seconds when he was resting?
   d. What is that in breaths per minute (60 seconds)?
   e. How much air went into his lungs each time he breathed when resting?
   f. Write two ways his breathing changed when he started running.

!

- Gas exchange means taking oxygen into the blood and releasing carbon dioxide.
- It happens in alveoli in the lungs.
- Respiration is the release of energy from food molecules such as glucose.
- It happens in all living cells.

Extension 10.3

Anaerobic respiration

Objectives
- Understand the process of anaerobic respiration
- Recognise that anaerobic respiration produces lactic acid

Muscles

Usain Bolt is fast. He won three Gold Medals in the 2012 Olympics. Marathon runners average 6 m/s, but Usain's average speed was more than 10 m/s, and he didn't take a breath until he crossed the finish line. His muscles made his feet push very hard against the track to move his body forward. That takes a lot of energy.

Usain's muscles could not get extra oxygen for respiration. So how did they contract enough to move him so fast? Sports scientists look for answers by studying runners' muscles.

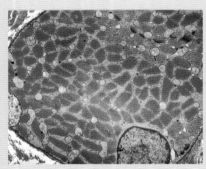

↑ This TEM image shows a slice through a muscle cell

Muscles cells have more than one nucleus. This microscope image shows two nuclei, both right at the edge of the cell, by the cell membrane. The cytoplasm is full of dark orange fibres. These use energy to make the muscle cell contract. The round shapes between the fibres are **mitochondria**. Most of the cell's respiration happens inside these.

Respiration

We usually use oxygen for respiration. The reaction shown in this equation shows what happens – it is called **aerobic respiration**. It releases all the energy stored in glucose molecules.

glucose + oxygen → carbon dioxide + water (aerobic respiration, energy)

Aerobic respiration can only release energy as fast as blood can bring oxygen to cells. Usain's races are over in seconds. He needs a lot of energy fast. His muscle cells store fuel but they can't store oxygen, and he can't wait for his blood to bring it.

When muscles work hard for less than 2 minutes they get most of their energy from a different method of respiration. It doesn't need oxygen so it is called **anaerobic respiration**.

Anaerobic respiration only releases 5% of the energy in glucose. The glucose isn't completely broken down. But cells can use a lot of glucose molecules at once, and 5% of their total energy is a lot.

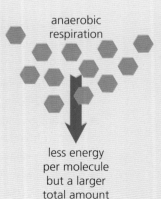

aerobic respiration — a lot of energy per molecule

anaerobic respiration — less energy per molecule but a larger total amount

↑ Anaerobic respiration can provide a quick burst of energy.

Respiration and breathing

Usain's muscles specialise in anaerobic respiration. All runners use anaerobic respiration when they start running, or speed up to cross the finish line. Anaerobic respiration lets cells release extra energy but it cannot be used for long periods.

Lactic acid

Anaerobic respiration has one major disadvantage. It makes lactic acid instead of carbon dioxide. Lactic acid is toxic if it builds up, so anaerobic respiration can't continue at full speed for more than a couple of minutes.

When a runner stops, they continue to breathe fast. The extra oxygen they take in after a race removes lactic acid and helps the muscles recover.

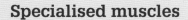

↑ After exercise, extra oxygen is used to break down lactic acid.

Specialised muscles

We all have two main types of muscles fibre – fast-twitch fibres and slow-twitch fibres. They are mixed together in your muscles but they behave very differently. Fast-twitch fibres produce large forces for a short period of time. They specialise in anaerobic respiration. Slow-twitch fibres contract more slowly and produce less force, but they can keep on contracting for a longer time. Scientists think people are born with different proportions of slow-twitch and fast-twitch muscle fibres.

Features	Muscle type	
	Slow-twitch	Fast-twitch
colour	dark	light
contraction speed	slow	fast
endurance	high	low
force produced	small	large
capillaries	many	few

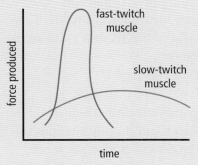

↑ Fast and slow twitch muscle fibres suit different sports.

Athlete	Calf muscle (%)	
	Slow-twitch	Fast-twitch
marathon runner	79	21
100 m sprint runner	24	76
weightlifter	44	56

Q

1. Heart muscle cells have a lot of mitochondria. Why do you think this is?
2. Make a table to summarise the differences between aerobic and anaerobic respiration.
3. Why is anaerobic respiration useful?
4. Use the data about fast-twitch and slow-twitch muscles to answer these questions:
   a. Which muscle type can exert more force?
   b. Which type can keep going for longer?
   c. What evidence suggests that slow-twitch fibres take up oxygen faster?
   d. What could account for a sprinter being able to run faster than a marathon runner?
5. Cheetahs are the fastest animals on land. What sort of muscle fibre would you expect to find more of in their leg muscles?

- Anaerobic respiration does not need oxygen.
- It is used during short bursts of vigorous exercise.
- It produces lactic acid, which is toxic.

10.4 Smoking and lung damage

Objectives
- Describe the effects of smoking
- Name some harmful substances in cigarette smoke

Smokers' lungs

Normal lungs are soft and pink, but these are a smoker's lungs. They are thick with **tar** and can't take in much oxygen. Smoke contains hundreds of different chemicals. They damage your lungs and increase your risk of getting heart disease or having a stroke. On average, smokers live 13 years less than non-smokers.

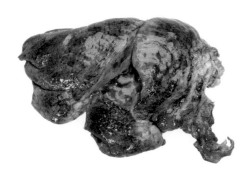

↑ Smokers' lungs are full of tar.

Keeping lungs clean

Healthy lungs can keep themselves clean. The main airways are lined with specialised cells. Some make sticky **mucus**. The rest are **ciliated**, which means covered in tiny waving hairs. The air you breathe in contains dust and micro-organisms. They get trapped in strands of mucus. Then the **cilia** sweep them up and out of your lungs.

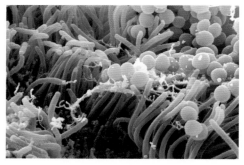

↑ The yellow micro-organisms are trapped in mucus. Ciliated cells will sweep them out of the lungs.

Tar

Tar paralyses cilia. Then dirt, microbes, and chemicals from cigarette smoke all build up in the mucus. That makes oxygen uptake much slower. Smokers also get more lung infections because microbes multiply in the mucus. The build-up of mucus irritates the lungs, so smokers try to cough it up. Eventually their airways become very narrow, making it hard for them to breathe. Their alveoli walls can also break down. They may need to breathe pure oxygen to stay alive.

Tar is also **carcinogenic**, which means it can make cells start to reproduce faster than they should and produce a lump called a **cancer**.

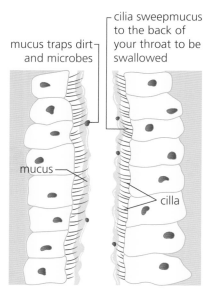

↑ Cigarette smoke destroys the cilia that keep lungs clean.

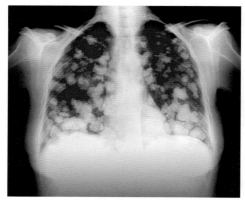

↑ The orange lumps in this chest X-ray are cancer cells growing out of control. They will eventually prevent the lungs from taking in oxygen.

Carbon monoxide

Cigarette smoke contains **carbon monoxide** – a toxic gas. It binds tightly to red blood cells. Then they can't carry so much oxygen. A smoker's heart tries to compensate. It beats more quickly and pumps more blood around. But if a smoker tries to run, they get out of breath very quickly.

When pregnant women smoke it also affects their baby. It increases the risk that the baby will be smaller than average when it is born.

Nicotine

Cigarettes are hard to give up once you've tried them. That means they are **addictive**. The **nicotine** in cigarette smoke affects brain cells. It makes them depend on getting nicotine to function normally. People who are addicted to cigarettes become irritable if they can't get one.

Nicotine also narrows blood vessels and increases blood pressure, which makes smokers more likely to suffer heart disease or a stroke.

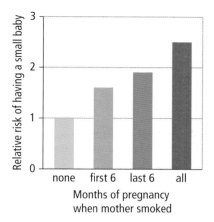

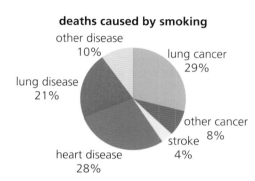

⬆ The main causes of death due to smoking.

Q

1 Copy and complete this table to show what sort of damage each chemical in cigarette smoke causes.

Chemical	Damage caused
tar	
carbon monoxide	
nicotine	

2 How is a baby affected if its mother smokes?

3 Does smoking do most damage at the start or end of pregnancy?

4 The diagrams below show alveoli from a non-smoker and a smoker.

non-smoker   smoker

⬆ Cigarette smoke breaks down the walls between alveoli.

a Describe what has happened to the smoker's alveoli.

b How will that affect their oxygen uptake?

5 Why do smokers find it hard to give up smoking?

!

- Cigarette smoke contains many harmful chemicals.
- Tar paralyses cilia and makes mucus build up in the lungs.
- Tar is carcinogenic so it can cause lung cancer.
- Carbon monoxide reduces the amount of oxygen blood can carry.
- Nicotine raises smokers' blood pressure and makes cigarettes addictive.

Enquiry 10.5 Communicating findings

Objectives
- Discuss results using scientific knowledge and understanding
- Communicate explanations clearly to others

Evidence

Kemi does not feel well. Her chest feels tight and she gets breathless if she tries to run. Her doctor thinks she may have **asthma**.

To collect more evidence, the doctor measures Kemi's **peak expiratory flow** (PEF). This shows how fast she can blow air out. Kemi has a peak flow of 320 dm³/minute. This result confirms his diagnosis. He asks his nurse, Joy, to explain what asthma means to Kemi.

Explanation

Joy knows that asthma is a chronic disease, which means it cannot be cured. Kemi will have the disease for a long time. She needs to understand what is wrong and use the medicines she is given properly.

Using a graph

Joy shows Kemi a graph. There is a wide range of normal PEF values. These values are affected by your age, gender, height, and health. The graph was made by testing a large number of students with each height and taking an average of their PEF values. Using a large sample size means that the average is more trustworthy.

The lines above and below the average show how much variation there is. 95% of students have PEF values between the two blue lines.

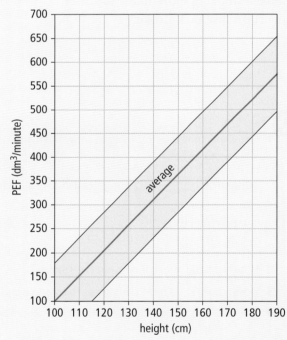

⬆ The graph shows how the average young person's PEF varies with height.

Kemi uses the graph to see how her result (320 dm³/minute) compares with the average for her height (160 cm). Graphs help us to visualise evidence and make it easier to spot patterns in the data. The PEF chart helps Kemi to see why she has been struggling to breathe. Now she wants to know what causes her symptoms.

Respiration and breathing

Using a model and images

Joy knows a lot about lungs but she plans to make her explanation simple. She will leave out anything that isn't completely relevant.

First she uses a straw as a model bronchiole. The straw is not much like a bronchiole but it helps Kemi to understand that air has to flow through tubes to get into her lungs. She struggles to breathe when the tubes in her lungs get too narrow. Models are useful because they make difficult ideas easier to understand.

Some substances irritate bronchioles. Joy uses pictures to explain what happens when bronchioles are irritated. They make extra mucus and try to protect the lungs by squeezing shut.

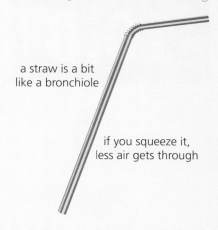
a straw is a bit like a bronchiole

if you squeeze it, less air gets through

In patients with asthma the bronchioles squeeze shut too easily. When too many bronchioles close up, they make it hard to breathe.

Joy shows Kemi how to use a blue inhaler when her chest feels tight. It relaxes the bronchioles so they don't close up so much. She advises Kemi to keep an asthma diary. She can learn to avoid substances that make her asthma worse.

Kemi rarely gets asthma attacks now. Joy's clear explanations made her understand how to take care of herself.

1. Is Kemi's PEF result higher or lower than the average value for her height?
2. What could be making Kemi's PEF result so different?
3. Draw a diagram to show the main differences between a normal bronchiole and one that belongs to someone with asthma.
4. Write what you would say to Kemi to explain her PEF value.
5. A patient has come to the hospital with chest pains. She is finding it hard to breathe. A scan shows a blood clot is blocking the main artery in one of her lungs. Write what you would say to convince the patient that she needs immediate treatment.
6. Damon doesn't smoke but he coughs a lot and finds it very hard to breathe. Tests show that the mucus in his lungs is very thick and sticky. His ciliated cells can't move it. The doctor gives him special medications to help thin the mucus. Write what you would say to convince him to use them properly.

- It is important to communicate findings clearly.
- Graphs, models, and pictures can make explanations clearer.
- Keep explanations simple by leaving out details that aren't relevant.

Review 10.6

1. The diagram below shows the main organ in the respiratory system.

   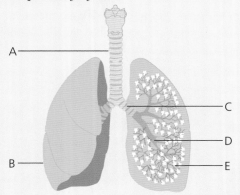

   Choose the correct letter for each of these parts:
   a. alveoli [1]
   b. bronchiole [1]
   c. trachea [1]
   d. lung [1]
   e. bronchus. [1]

2. Name the part of the lungs that matches each of these descriptions:
   a. small air spaces where gas exchange takes place [1]
   b. a reinforced tube that takes air down to one lung [1]
   c. small tubes that lead air to millions of tiny air spaces. [1]

3. The model lung in the diagram can be used to show how air is moved in and out of the lungs.

   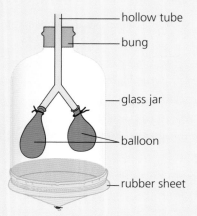

   a. Which two parts of the lungs does the hollow tube represent? [2]
   b. What does the rubber sheet represent? [1]
   c. Describe what happens when the rubber sheet is pulled down. [1]
   d. A model is never exactly like the real thing. Name one part of the lungs that is not represented on this model. [1]

4. When your chest expands, air moves into your lungs. Write the letters A–E in the correct order to describe the path the air takes. [2]
   A alveoli
   B bronchi
   C trachea
   D bronchioles
   E nose

5. The diagram below shows how gas exchange takes place in the lungs.

   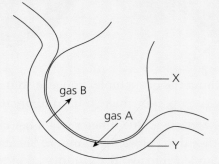

   a. Name parts X and Y. [2]
   b. Name gases A and B. [2]
   c. X and Y have very thin walls. Explain how this helps gas exchange. [1]

6. A student walked for 10 minutes and ran for 10 minutes. After each activity she checked her breathing rate and her heart rate.

Activity	Breaths per minute	Heart beats per minute
walking	12	60
running	36	140

   a. How were her heart and lungs affected when her muscles worked harder? [1]
   b. Explain why both organs need to respond in the same way. [2]

7. Processes A, B, and C all involve oxygen and carbon dioxide.
   A gas exchange
   B breathing
   C respiration

Choose the correct letter to match each of these descriptions:

a moving air in and out of the lungs [1]

b the process that takes place in alveoli [1]

c a chemical reaction that takes place in every cell. [1]

8 The apparatus in the diagram collects gases released by a burning cigarette.

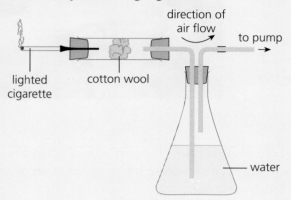

a When a cigarette burns the cotton wool turns black and sticky. Name the black substance produced. [1]

b The cotton wool in the apparatus represents the inside of a smoker's lungs. Describe one effect the black substance can have on lungs. [1]

c When universal indicator is added to the water it turns red. What does this tell you about cigarette smoke? [1]

d One of the chemicals in cigarette smoke is nicotine. How does this chemical make it hard for people to give up smoking? [1]

9 Two types of cell line the main airways in the lungs: cells that produce a sticky mucus, and ciliated cells like the one below.

The dust and micro-organisms in air stick to the mucus in the lungs.

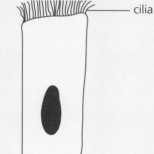

a What do ciliated cells do to mucus? [1]

b The smoke from cigarettes destroys cilia. How can this cause more lung damage? [1]

10 Aerobic respiration requires oxygen.

a Copy and complete the equation for the reaction. [1]

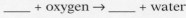
____ + oxygen → ____ + water

b Explain why respiration needs to take place in every cell. [1]

11 10 smokers and 10 non-smokers take part in an investigation. They are all aged 15. Each takes one breath and blows into a balloon as hard as possible. Their average results are shown in the bar chart.

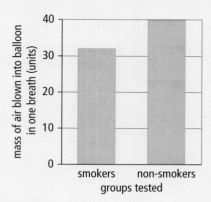

a What is measured in this investigation? [1]

b Write down one factor that is kept the same. [1]

c Do the results support the idea that smoking damages lungs? [1]

12 A student walks for 10 seconds, runs for 10 seconds and then walks again. A machine plots a graph to show how the volume of air in his lungs changes.

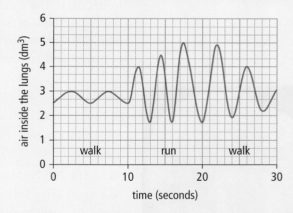

a What is the largest volume of air in his lungs when he starts to walk? [1]

b What is the largest volume of air in his lungs while he runs? [1]

c Calculate how many breaths he takes per minute (60 seconds) when he starts to walk. [1]

d Calculate how many breaths he takes per minute while he runs. [1]

e Compare his breathing while he is walking and while he is running. Write down two differences. [2]

11.1 Reproduction

Objectives
- Describe the human reproductive system
- Understand what happens during fertilisation

Fertilisation

Akanni's life began like everybody else's. A **sperm cell** from his dad joined with one of his mother's **egg cells**. This is **conception**. Joining an egg and sperm to start a new life is called **fertilisation**.

The picture below shows two sex cells – an egg and a tiny sperm cell. The egg cell is about the size of this full stop.

Eggs

A woman's eggs are all formed before she is born. They are stored in organs called **ovaries**. One egg usually matures each month. It bursts out of an ovary and gets swept along a tube called an **oviduct**.

Egg cells can't move, but the cells that line the oviduct are coated with tiny cilia, like the ones in your airways. These sweep the egg down towards the **uterus**. It can survive for 24 hours, but it dies if it isn't fertilised by then.

When a man and woman make love, sperm enter the woman's **vagina**, which leads up to her uterus. The sperm can only survive for a day or two. They need to get up to the oviduct to fertilise the egg before it's too late. If fertilisation happens, the egg forms an **embryo** which settles in its mother's uterus and begins to grow. Then 9 months later a baby is born. Muscles in the uterus contract and squeeze it out through the vagina.

Sperm

Sperm are cells with a mission. They swim, find an egg, and fertilise it. About 500 million sperm land in a woman's vagina, but each egg only lets one sperm in. The rest die.

Akanni's dad has been making millions of sperm every day since he was a teenager. They form in his **testes**. When he makes love, they get pumped out along his **sperm ducts**. They pass glands which add fluids to help the sperm stay alive. Then the mixture of sperm and fluid – called **semen** – spurts out through his **penis**.

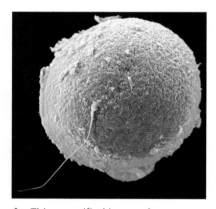

↑ This magnified image shows a sperm fertilising an egg

Reproduction and fetal development

Sex cells

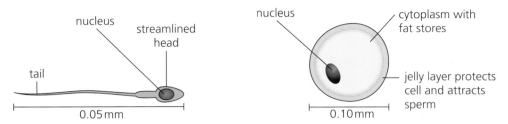

↑ Sperm and egg cells are specialised to suit their functions.

Sperm and eggs are specialised for the functions they carry out. A sperm cell is a fast swimmer and it can release an enzyme to help it penetrate the egg.

Each egg contains a large store of food. When it is fertilised it uses the food to produce an embryo. An egg is also very well protected. As soon as one sperm enters an egg, its outer membrane seals. So only one sperm gets in. Its nucleus fuses with the egg cell nucleus and a new life begins.

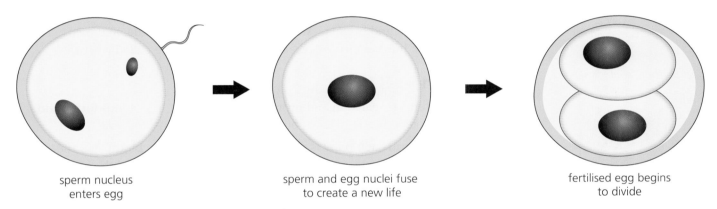

↑ A new life is created when sperm and egg nuclei fuse.

A fertilised egg like this was Akanni's first cell. It began to divide to make more cells and build an embryo.

Q

1. What happens during fertilisation?
2. Copy and complete this flow chart to show the path eggs take when they leave the ovary.

   ovary → _____ → _____
3. How often is an egg released in a woman's body?
4. Now copy and complete this flow chart show the path sperm take when they leave the testes.

   testes → _____ → _____ → vagina → _____ → _____
5. Where does fertilisation usually happen?
6. What happens to eggs and sperm if they don't meet?
7. What is a baby called when it first starts to grow?
8. What features make sperm cells good at swimming?
9. How are egg cells suited to their function?

!

- Fertilisation happens when a sperm nucleus fuses with an egg cell nucleus.
- The egg is released from an ovary and swept along the oviduct to the uterus.
- Sperm are made in the testes, and pumped out through the sperm duct and penis.
- Sperm swim from the vagina, through the uterus, to the oviduct to find the egg.

137

11.2 Fetal development

Growing

You began life as a single, fertilised egg cell. This cell split into two, and each new cell divided again and again. A week later you arrived in the uterus. Your cells had divided eight times and formed a hollow ball. The ball of cells settled into the wall of the uterus. This is **implantation**.

A growing body needs food and oxygen. When you settled in the uterus, some of your cells made contact with your mother's blood supply. They began to form a placenta and umbilical cord to pass nutrients and oxygen from her blood to yours.

The rest of your cells carried on dividing. They started to specialise and produce different tissues and organs – starting with your brain, spine, and heart. You began to look human. After 8 weeks of growth you were called a **fetus**.

Safe and well fed

A growing fetus is sealed in and protected. It floats in a bag of **amniotic fluid**. The fluid protects the fetus from bumps and lets it practise moving.

Each mother and each baby is unique, so they can't share blood. But the **placenta** lets molecules move between the mother's blood and the blood of the fetus. It does the jobs the small intestine and lungs will do when the fetus is born. But it takes food molecules and oxygen from the mother's blood instead of the outside world.

Human blood is always red, but on the diagram the blood from the fetus is coloured blue. You can see how close it comes to the mother's blood. The blood vessels in the placenta have a large surface area, so that nutrients and oxygen can diffuse into them quickly. At the same time waste molecules like carbon dioxide diffuse from the fetus to the mother.

Objective
- Describe fetal development

Day 1
fertilisation

Day 2
cells divide

Day 3

Day 4

Day 6
hollow ball of cells forms

Day 7
embryo implants in the wall of the uterus

Day 28
placenta begins to take nutrients and oxygen from mother's blood

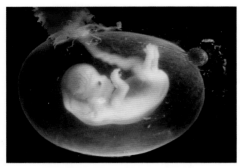
↑ Safe and well fed – an 8-week-old fetus.

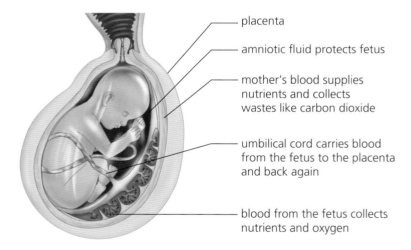
- placenta
- amniotic fluid protects fetus
- mother's blood supplies nutrients and collects wastes like carbon dioxide
- umbilical cord carries blood from the fetus to the placenta and back again
- blood from the fetus collects nutrients and oxygen

Danger

A fetus is safe from physical damage, but not from chemicals and micro-organisms. Drugs and infections can cross the placenta. They can harm the fetus, especially when their organs are developing during the first 12 weeks. Pregnant women should not smoke or drink alcohol.

Becoming human

Just a dot	o	1 week – cells beginning to specialise
3 mm long		4 weeks – spine and brain forming, heart beating
3 cm long		9 weeks – tiny movements, lips and cheeks sense touch, eyes and ears forming
7 cm long		12 weeks – fetus uses its muscles to kick, suck, swallow, and practise breathing

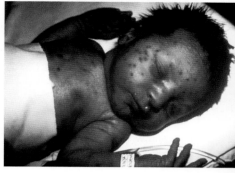

↑ This baby's brain didn't develop properly. She is deaf and blind. Her mother caught rubella early in her pregnancy and the infection damaged her fetus.

As you grew in your mother's uterus, your mind began to form too. Your brain, spine, and nerves were some of the first parts to grow. When your muscles formed, your brain and nerves made them work. You practised all the movements you would make when you were born.

By the twentieth week, most babies have a full set of sense organs – eyes, ears, touch, smell, and taste. These start to send signals to your brain about the world around you. At first your brain can't make sense of all the signals. Scientists think brains turn on gradually – like lights controlled by dimmer switches. It's a two-way process. Signals from your sense organs help your brain cells to connect with each other.

Q

1. When does an embryo implant in the wall of the uterus?
2. If a cell just kept dividing, it would make a big round blob. What else happens as the embryo turns into a fetus?
3. Write one function of the bag of fluid around the fetus.
4. How does a fetus get nutrients and oxygen?
5. Suggest why pregnant women can't take some medicines.
6. Women can take 6–8 weeks to realise they are pregnant. What problems might result if a woman does not know she is pregnant?

- Fertilised egg cells divide and specialise to form a fetus.
- The placenta transfers nutrients and oxygen from the mother's blood to the fetus, and removes waste.
- Amniotic fluid protects the baby from physical damage.
- A fetus can be damaged by cigarette smoke, alcohol, or micro-organisms.

Extension 11.3

Twins

Identical and non-identical twins

Babies sometimes arrive in twos or threes. This can happen when an ovary releases extra eggs and they all get fertilised, forming two or more embryos. Most twins and triplets are formed like this. They are no more alike than ordinary brothers or sisters.

Identical twins form if a single embryo splits in two very early in its development. Each half of the embryo grows into a whole new baby. Each baby gets its genes from the nucleus of the fertilised egg. So the cells in each baby follow the same instructions.

The split that makes identical twins usually happens between days 4 and 9. After 12 days it's too late for the two halves of an embryo to separate.

Objectives

- Understand how twins form and why some are identical
- Appreciate the extra risks twin babies face

↑ These twins are non-identical because they were formed from two different fertilised egg/s.

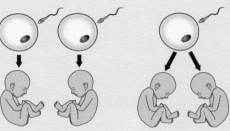

non-identical twins formed from two fertilised eggs

identical twins formed from one fertilised egg when the embryo splits

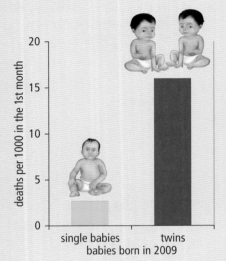

deaths per 1000 in the 1st month — single babies / twins — babies born in 2009

Problems

In 2009, the Office of National Statistics compared single babies with twins. Babies who shared a uterus were far more likely to die in the first month.

But why? What could be different about them?

percentage of births — single babies / twins — birth mass (kg): <1, 1–1.4, 1.5–1.9, 2–2.4, 2.5–2.9, 3–3.4, 3.5–3.9, >4

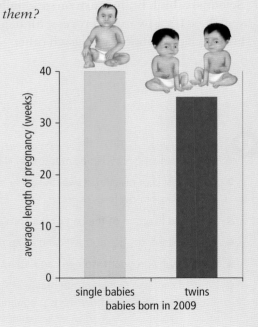

average length of pregnancy (weeks) — single babies / twins — babies born in 2009

Reproduction and fetal development

To find out more they compared the birth masses of single babies and twins. On average the babies who were twins had a lower birth mass. This raised another question.

Could the birth mass of babies affect their survival?

To answer this question they looked at the death rates of babies with a birth mass below 1 kg. About one-third of these babies died. It didn't matter whether they were twins or single babies. Most of the babies with very low birth weights were twins.

Why? Could it be because they were born too soon?

Doctors compared the pregnancy lengths of twins and single babies. On average twins were born earlier. This could account for their lower average birth mass and survival rate.

Despite the extra risks, twins born as early as 24 weeks have survived.

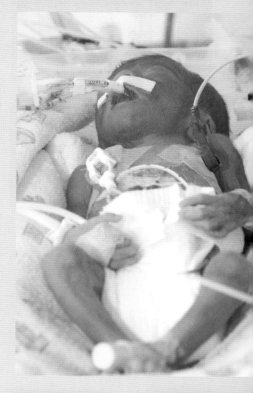

Premature babies

In a modern hospital, tiny babies can be kept alive. That's what all parents want. They need lots of expensive equipment and well-trained nurses. They cannot suck, so they need to be fed through a tube. Their lungs aren't ready yet either, so they need a machine to help them breathe.

Conjoined twins

A partially split embryo forms conjoined twins. They can be joined by their heads, chest, or back and sometimes they share internal organs. In very rare cases they share nearly everything except their heads. Unfortunately most conjoined twins don't survive.

If the babies don't share too many organs doctors try to separate them. The operations are very complicated. They often take more than 12 hours. An international team of experts may need to come to the hospital to help.

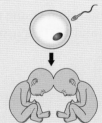

conjoined twins formed from one fertilised egg when the embryo does not split completely

---

1. Draw diagrams to show how identical and non-identical twins form.
2. Yusra had three babies – two identical girls and a boy. How could that happen?
3. Use the bar charts on this page to answer these questions.
   a. How many twins died in the first month per 1000 born?
   b. What is different about the birth masses of twins and single babies?
   c. How much longer is the average pregnancy for a single baby?
4. Some babies have problems with their vision or hearing. A hospital compared babies with low and high birth masses. What do their data show?

Birth mass (kg)	Poor vision (%)	Hearing problems (%)
less than 1	50	46
more than 3	1	1

5. Explain why premature babies need expensive medical care.

- Non-identical twins form when two sperm fertilise two eggs.
- Identical twins form when embryos split into two.
- Conjoined twins form when identical twins do not separate completely.

11.4 Adolescence

Puberty

Objectives
- Recognise the changes caused by puberty
- Understand why girls have periods

Jasmine and Rakesh were the same height last year. Now she is taller than him. She has had a **growth spurt**. When Rakesh reaches **puberty** he will start to grow faster too. His appearance and emotions will also change.

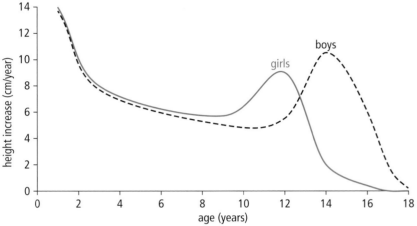

↑ Girls usually have their growth spurts earlier than boys.

Puberty is the time when young people mature. Puberty most often occurs between ages 11 and 15. It is usually earlier for girls than boys.

Hormones

Boys	Both	Girls
penis develops	hair grows under arms and in pubic region	breasts develop
voice deepens	growth spurt	periods start
hair grows on face	hair grows on legs	hips widen
sperm produced		uterus develops

↑ Some of the changes sex hormones cause are the same for boys and girls.

When you reach puberty, your brain causes your testes or ovaries to pour **hormones** into your blood. Hormones are chemical messengers which can affect many parts of your body. You have hormones in your blood all the time, but you only really notice your **sex hormones**. During puberty they cause the changes shown in the table.

Periods

A girl's **periods** start during puberty. She loses some blood through her vagina for 3–4 days a month. The bleeding is part of a series of events called the **menstrual cycle**.

Every month, hormones make an egg mature in one of her ovaries. At the same time her uterus builds up a thick lining. Then the egg is released. This is **ovulation**.

The egg travels slowly down the oviduct towards the uterus. If it is fertilised, an embryo forms and implants in the thick lining. If the egg isn't fertilised, the girl's hormone levels drop. This makes the lining of the uterus break down.

It leaves her body as a flow of blood through her vagina. We count the days of a menstrual cycle from the first day of this bleeding.

A menstrual cycle usually lasts 28 days, but can be longer or shorter. It can also vary from month to month.

Periods stop while a woman is pregnant and end completely at the **menopause**, when she is about 50 years old.

Feeling different

During puberty your sex hormones affect your mind as well as your body. They can make you feel moody and irritable. They can also upset your body's timer, so it is harder to get enough sleep. It feels like 8 p.m. when it's really 11 p.m. and it feels like 4 o'clock in the morning when it's time to get up for school.

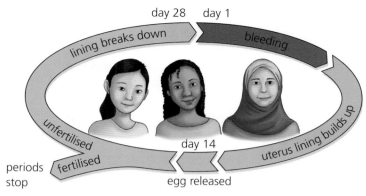

↑ A woman's menstrual cycle prepares her uterus to receive a fertilised egg.

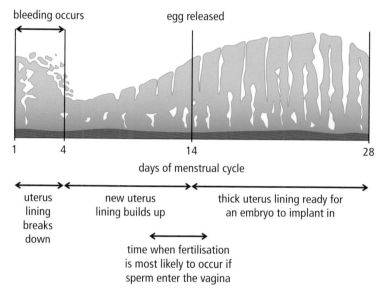

↑ A woman's menstrual cycle shows when fertilisation is most likely to occur.

Q

1. Which of the seven life processes does puberty prepare you for?
2. What makes puberty start?
3. Explain what hormones are.
4. Why do so many things change at once during puberty?
5. What makes a girl notice her periods have started?
6. Make a flow chart to show the events in the menstrual cycle. Start with: An egg begins to mature in one of the ovaries.
7. Why do some young people feel very tired during puberty?

- During puberty, sex hormones cause physical and emotional changes.
- Height increases rapidly, sex organs develop, and periods start in girls.

Review 11.5

1 The diagram shows two specialised cells.

  a Name cell A. [1]
  b Name the organs where these cells are produced. [1]
  c Name cell B. [1]
  d Name the organs where these cells are produced. [1]

2 The diagram shows part of human reproduction.

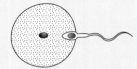

  a Name the process taking place. [1]
  b Name the place where this is most likely to happen. [1]

3 Name the parts of the male reproductive system that have each of these functions:
  a produces sperm [1]
  b ejects the sperm. [1]

4 Name the part of the female reproductive system that has each of these functions:
  a produces eggs [1]
  b collects sperm [1]
  c holds the fetus as it develops [1]
  d provides a passage for a baby to squeeze through when it is born. [1]

5 An egg only fuses (joins) with one sperm. Write the letters in order to describe how this happens. [2]
  A The fertilised egg cell begins to divide to produce an embryo.
  B The egg stops any more sperm cells getting in.
  C Many sperm cells swim to the egg.
  D The successful sperm's nucleus fuses with the egg cell's nucleus.
  E One sperm gets into the egg cell.

6 During puberty, sex hormones cause physical changes. For each of the following changes, write whether it happens in boys, girls, or both: [4]
  a voice gets deeper
  b hips get wider
  c hair grows in the pubic area
  d breasts develop.

7 The chart shows how the lining of the uterus changes during a typical month.

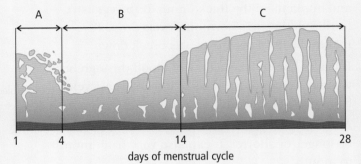

days of menstrual cycle

Choose the correct letter for each of these parts of the cycle.
  a when the uterus lining gets thicker [1]
  b when the uterus lining breaks down [1]
  c when the lining stays thick to receive a fertilised egg [1]
  d when bleeding takes place [1]

8 A sperm cell is specialised so that it can find and fertilise an egg cell.

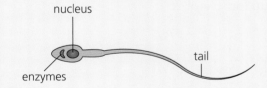

  a Describe one feature that helps it move towards the egg. [1]
  b Describe one feature that helps it penetrate the egg. [1]
  c Which part of the sperm cell fuses with the egg to fertilise it? [1]

Reproduction and fetal development

9 For fertilisation to happen, a sperm needs to meet an egg soon after it is released from the ovary.

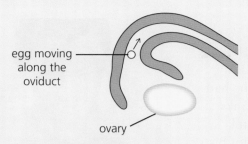

a On which day of the menstrual cycle is the egg usually released? [1]

b On which day of the menstrual cycle does bleeding begin? [1]

c A woman with a regular 28-day cycle begins her period on 5 May. On which date is she likely to ovulate? [1]

10 The diagram shows a woman's uterus when she is close to giving birth.

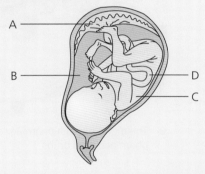

Write the correct letter for each of these parts:

a amniotic fluid [1]
b fetus [1]
c umbilical cord [1]
d placenta. [1]
e What is the function of liquid B? [1]
f What job does part D do? [1]

11 The graph shows how the average mass of a fetus changes during pregnancy.

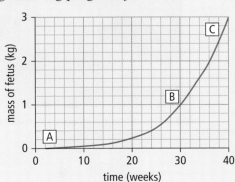

Choose the correct letter for each of these stages of pregnancy:

a when the mass of the fetus is increasing most rapidly [1]

b when it might be possible to keep the fetus alive if it was born early [1]

c when the fertilised egg divides to form an embryo and implants in the wall of the uterus. [1]

12 The placenta allows small molecules to move from a mother's blood to the blood of the fetus.

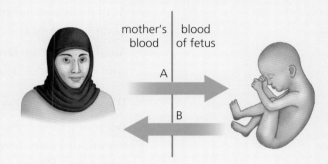

Write the letter of the correct arrow to show which way each substance moves:

a food molecules [1]
b carbon dioxide [1]
c oxygen [1]
d alcohol. [1]

13 Doctors compared babies whose mothers smoked with babies whose mothers did not. These are the lines of best fit from their graphs.

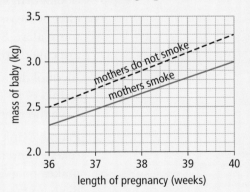

a Many of the babies were born early. How does the mass of the baby change as the pregnancy gets longer? [1]

b Scientists think cigarette smoke can damage a growing fetus. Does this evidence support that idea? Explain why. [2]

c Name another substance that can damage a growing fetus. [1]

12.1 Drugs

Objective
- Recognise how conception, growth, development, behaviour, and health can be affected by drugs

Alcohol

Arjun has a terrible headache. He drank alcohol for the first time last night. His parents are away so he had friends round. They brought drinks with them. Arjun had never drunk alcohol before, but he didn't want his friends to know that. He drank a lot. At first the alcohol made him relax. Then he was sick.

Alcohol can leave drinkers with serious headaches.

Alcohol is a drug. **Drugs** are chemicals that affect your cells and change the way your body works. There are two sorts: pharmaceutical drugs and social drugs.

Pharmaceutical drugs

Doctors use pharmaceutical drugs to cure diseases or reduce their symptoms. Pharmaceutical drugs include painkillers such as aspirin, and antibiotics such as penicillin. Doctors use antibiotics to treat infections caused by bacteria.

Most drugs needed to be treated with care. They can have harmful side effects if you take too much.

The medicines in a pharmacy contain pharmaceutical drugs.

Social drugs

Social drugs affect the nervous system. People use them to experience the effects the drugs have on their brain. The nicotine in cigarettes makes people more alert. So does the **caffeine** in coffee, tea, and fizzy drinks. Drugs like this are called **stimulants**.

Alcohol does the opposite. It slows your reactions so it is a **depressant**. Drinkers take longer to spot hazards, so they should not drive. They would have too many accidents.

People drink alcohol to feel more relaxed and sociable. But alcohol also makes people aggressive and more likely to take risks. Drinkers often get into fights and injure themselves.

Most drugs have unpleasant side effects. Cigarettes damage your heart and lungs. Alcohol damages your liver and increases your risk of heart disease. Caffeine can cause depression and keep people awake at night.

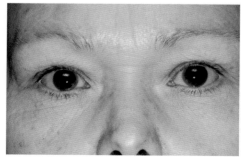

This drinker's yellow eyes show that her liver isn't working.

Illegal drugs

The drugs with the most serious side effects are illegal. They include depressants such as heroin and morphine, and stimulants such as cocaine and amphetamines. Taking any of these can be fatal.

Cannabis

Cannabis is also called Indian hemp and hashish. It changes the way users see and hear things, so it's a **hallucinogen**. It makes people who use it relaxed and talkative, but it can also cause panic attacks. Some users become addicted and some develop mental health problems and become infertile. The smoke from burning cannabis is full of toxins, like tobacco smoke (see page 130), so it also causes lung damage.

↑ All of these drugs contain cannabis.

Damaged for life

Emmanuel has a poor memory and he finds it difficult to concentrate in school. He has **fetal alcohol syndrome**. His mother drank a lot before she realised she was pregnant.

High alcohol levels in the blood can damage embryos. They usually survive, but they are often smaller than average when they are born. When they grow up they are likely to have learning difficulties or mental health problems.

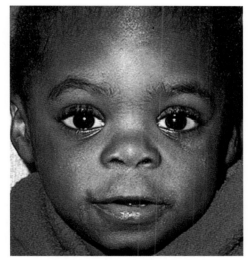

↑ Emmanuel's brain was damaged by alcohol when he was an embryo.

Q

1 Name one legal social drug that has each of these effects:
   a makes you feel alert
   b relaxes you.
2 Name the main organs that the following drugs damage:
   a cannabis
   b alcohol (in adults)
   c alcohol (in developing embryos).
3 Cannabis makes nerves send distorted messages. What effect does that have on the people who use it?
4 Joshua never drinks alcohol before a football match. He says it would make him a worse player. Explain why he is correct.

!

- Pharmaceutical drugs are used as medicines.
- Social drugs are used for the effects they have on the brain.
- Drugs can make you more alert, relax you, or change the way you see and hear things.
- Illegal drugs have serious side effects and using them can be fatal.

12.2 Disease

Tropical diseases

Objectives

- Recognise how conception, growth, development, and health can be affected by disease

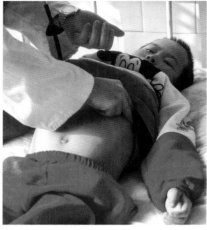

Chao has bilharzia. He caught it when he played in an infected lake. Worms have been living in his blood ever since. Their eggs leave his body in his urine and faeces. The disease has given him anaemia and diarrhoea and is damaging his liver. He is very small for his age.

↑ Chao's swollen belly is caused by worms. They are reproducing in the blood vessels around his alimentary canal.

↑ Worms like these cause bilharzia.

↓ These worms all came from one small boy's intestines.

↑ These children's parents have river blindness caused by a worm carried by blackflies.

↑ Worm larvae like these cause river blindness when they infect eyes.

Olawale and Khadijah don't go to school. Their parents have river blindness. Infected blackflies bit them. Now thousands of tiny worms have spread through their bodies. The worms made them blind and made their skin very itchy. Their children need to stay at home to help them.

Diseases like bilharzia and river blindness disrupt students' education, damage their health, and slow down their growth. They are called neglected tropical diseases because not much money has been spent on treating victims, or stopping the diseases spreading.

Worms

This huge mass of worms was extracted from one small boy. The worms live and reproduce in their victim's small intestine. Their eggs escape in faeces. More children become infected by eating food contaminated with the eggs. The worms slow growth, cause malnutrition and anaemia, and make it harder for students to do well in school.

Tuberculosis

Priya can't stop coughing. She has a fever and she is very thin. Her chest X-ray confirms that she has tuberculosis (TB). The bacterium multiplies in victims' lungs and spreads through the air when they cough. TB is an unusual disease though. Only 1 in 10 of those infected become ill, so people with no symptoms can still pass on the disease. It is very difficult to cure.

↑ TB is very hard to cure.

↑ The cloudy white areas on this X-ray show that this woman has TB.

Sexually transmitted infections

The microbes that cause **sexually transmitted infections** (STIs) are passed on during sex. They can only survive for a few seconds outside the body. **Condoms** protect people from most STIs if they are used properly.

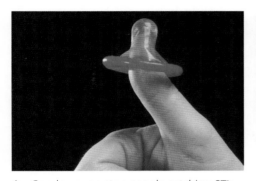

↑ Condoms can stop people catching STIs.

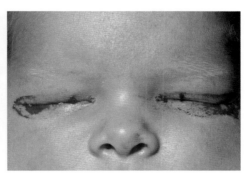

↑ Babies can pick up STI's while they are being born.

This baby's mother had no symptoms. She didn't know she had **chlamydia** and **gonorrhoea** – the two most common STIs. These bacteria infect cells around the entrance to a woman's uterus. The bacteria got into her baby's eyes and lungs as she was giving birth.

STIs also affect conception and many women with these diseases can't get pregnant. STIs can cause **infertility** in both men and women.

Q

1 Make a table like this to show a disease you could catch in each situation.

Situation	Disease that could be caught
walking into a pond in Africa, South-East Asia, or the Middle East with bare legs	
living near a stream in Africa	
eating contaminated food in a tropical country	
sleeping in a crowded room where there may be carriers of the disease	
having sex without using a condom	

2 Why is it a problem when an infected person has no symptoms?

3 Suggest two reasons why infectious diseases stop families escaping from poverty.

- Infectious diseases damage health, slow growth, and disrupt student's education.
- STIs can cause infertility or be passed to a baby during birth.

Extension 12.3: Defence against disease

Objective
- Recognise that the immune system destroys pathogens

Pathogens

It's hard to avoid micro-organisms. You breathe them in, swallow them in your food and drink, and pick them up when you touch things. The harmful ones are called **pathogens**. They cause infectious diseases.

One of the biggest threats to human health is untreated **sewage**. This is a mixture of dirty water and human faeces. In modern cities, water is piped to homes and sewage is taken away. But in other places, bacteria from faeces can get into drinking water – from dirty hands or soil. This can spread diarrhoea, typhoid, and cholera to thousands of other people.

⬆ Dirty water spreads many diseases.

Keeping pathogens out

Your skin keeps most pathogens out. Gaps in the skin have extra defences. Stomach acid keeps most microbes out of your alimentary canal and mucus traps most of the pathogens you breathe in.

Mosquitoes get round these defences by injecting pathogens straight through your skin into your blood. Dengue, yellow fever, and malaria are all caught from mosquitoes.

Destroying invaders

If pathogens do invade your tissues, your **immune** system can usually destroy them. It uses several types of white blood cell to do this. Some types can detect pathogens and destroy them. They are **phagocytes**. This one is surrounding the bacteria (pink) with its cytoplasm. It will pull them in and digest them.

Eventually, these white cells die. The green pus that oozes from wounds is full of their broken-down cells.

⬆ This mosquito sucked blood from a dengue victim. She will transmit the virus to anyone else she bites.

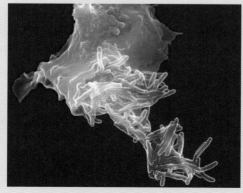

⬆ This phagocyte will wrap the pink bacteria in its cytoplasm and digest them.

Drugs and disease

Antibodies

If phagocytes can't deal with an infection, other white cells come to your defence. They make **antibodies**. These Y-shaped proteins circulate in your blood until they find a pathogen. Then they stick to it. Some antibodies destroy pathogens themselves. Other antibodies help white blood cells to locate and destroy the pathogens.

It takes time to produce antibodies, so pathogens can make you feel ill for a few days before your immune system destroys them.

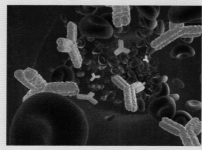

↑ These Y-shaped antibodies circulate in the blood until they find the pathogen they recognise.

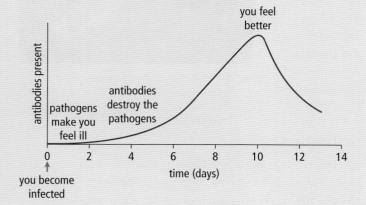

Some micro-organisms evade your immune system. Malaria protozoa and chlamydia bacteria hide inside human cells. That makes them harder to destroy. Other pathogens can prevent antibodies from recognising them. They change the outside of their cells, or disguise themselves by wrapping themselves in human proteins.

↓ An HIV virus turned this cell into a virus-making factory. Now new virus particles are escaping from its surface.

AIDS

Some white blood cells can be disabled by viruses. This one is dying. HIV viruses infected it. They turned it into a virus-making factory. Now the new red virus particles are escaping. They will circulate in their victim's blood until they find another cell like this to destroy.

After many years, HIV can weaken a victim's immune system. Then the disease turns into **AIDS** (acquired immune deficiency syndrome). AIDS makes it harder to fight off other pathogens, especially the ones that cause pneumonia and TB.

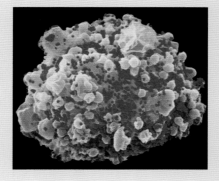

Q

1. List two ways microbes can enter your body.
2. Cholera bacteria cause severe diarrhoea. These bacteria spread rapidly when sewage is not piped away. Why is this?
3. How are bacteria kept out of your alimentary canal and lungs?
4. Describe how phagocytes destroy pathogens.
5. How do antibodies help phagocytes?
6. What is the connection between HIV and AIDS?
7. Why do people with AIDS often die from other infections?

- Your skin keeps most microbes out.
- Harmful micro-organisms are called pathogens.
- Pathogens that invade your body are attacked by white blood cells.
- Some white blood cells make antibodies to help destroy pathogens.

Extension 12.4: Boosting your immunity

Objective
- Recognise how conception, growth, development, behaviour, and health can be affected by disease

Pathogens

Pathogens can use your tissues as food, and fill your body with toxic waste. One or two micro-organisms can't do much harm, but they grow and reproduce very quickly. They can double their numbers in less than half an hour. So 1 becomes 2, 2 become 4, and very soon there are millions. Your immune system needs to destroy them before their numbers get too large.

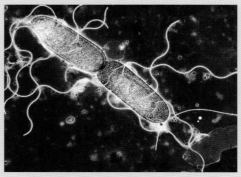

↑ Bacteria reproduce by dividing into two. They can be ready to divide again in less than half an hour. Bacteria like this cause food poisoning.

↑ Antibodies take time to produce.

Antibodies

Every micro-organism is different. Each set of antibodies is specially made to fit one pathogen. The first time a new pathogen infects you, white cells take about a week to make the right antibodies. This delay gives the pathogen time to build up its numbers. The pathogens can make you feel very ill, but your immune system usually destroys them eventually.

Antibodies only last a few days, but white blood cells remember how to make them. If the same pathogen returns, they make these antibodies straight away. So the pathogen is destroyed before it has time to reproduce, and you don't get ill. You are immune to the disease it causes.

Vaccination

This child is being **immunised** – made immune to a disease. The syringe contains a **vaccine**.

Vaccines are weak or damaged pathogens. They can't make you ill, but your white cells still make antibodies to fit them.

If the real pathogen infects you later, you don't notice it. Your white cells produce antibodies very quickly, just as if it was a second infection. They destroy the pathogen before it can harm you.

↑ Vaccination makes you immune to specific diseases.

Drugs and disease

Smallpox used to kill more people than any other disease. It was wiped out in the 1970s by immunising millions of people. The last person to catch the disease was very foolish. He was scared of needles, so he pretended he'd already been vaccinated. He was lucky to survive.

Now the vaccine is no longer needed. This pathogen has been destroyed. So many people were vaccinated that there was no one left for it to infect.

Vaccines are hard to make. It can take 20 years and cost billions to develop a new vaccine. We still don't have vaccines for many common diseases.

Antibiotics

Sometimes white blood cells can't cope with an infection. Bacteria continue to multiply in your tissues. If you are very ill, an antibiotic could save you. They destroy bacteria, but not other pathogens.

This plate looks cloudy because there are bacteria growing on it. The white tablets are antibiotics. They have destroyed all the bacteria in the clear circles around them. Some bacteria are **resistant** to antibiotics – they are not destroyed. The smaller the clear circle, the more resistant to the antibiotic the bacteria are.

If you are given antibiotics, you must finish all the tablets, even if you feel better. The toughest bacteria are the last to be killed. If these survive and multiply, they will be harder to destroy.

When old antibiotics stopped working, scientists used to make new ones. Now they are running out of new ideas to try. Some types of TB are incurable because the bacteria that cause them can resist every antibiotic we have.

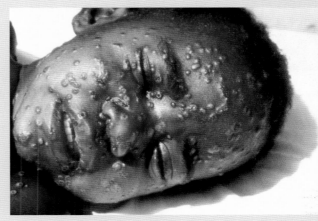
↑ The smallpox virus was the first pathogen to be wiped out completely by immunisation.

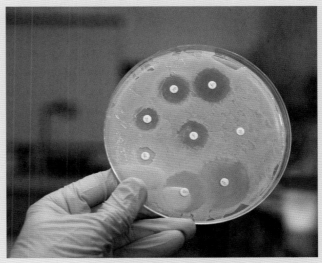
↑ The clear circles show where antibiotics have destroyed the bacteria on this plate.

Q
1. How do bacteria reproduce?
2. Why do most pathogens take a day or two to make you ill?
3. How do you become immune to a disease?
4. What do vaccines contain?
5. How does immunisation work?
6. Why can't antibiotics cure a cold or flu?
7. Hospitals often test different antibiotics on the pathogens from patients' wounds. Suggest why.
8. How do bacteria become resistant to antibiotics?

- Pathogens reproduce very quickly.
- White blood cells produce different antibodies for each pathogen.
- Vaccines contain harmless pathogens that make white blood cells produce antibodies.
- A first infection, or vaccination, makes you immune to a pathogen, because antibodies against it can be made more quickly.

Review 12.5

1. A, B, and C are organs that can be damaged by drugs.

   A heart

   B lungs

   C liver

   Choose the correct letter to match each of these descriptions:

   a two organs damaged by smoking [2]

   b two organs are damaged by alcohol. [2]

2. A survey was carried out into the drinking habits of men and women in the UK.

Age	% who drink more than the recommended amount	
	Men	Women
Under 35	39	21
Over 35	35	16

   Write two conclusions that can be made from these results. [2]

3. Chenglie tests his reaction speed by asking a friend to drop a ruler and recording how far the ruler drops before he catches it. He repeats the test after taking some caffeine.

   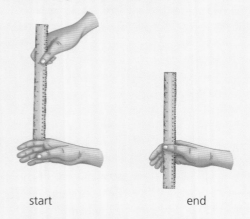

   start   end

Test	Distance the ruler drops (cm)	
	Before caffeine	After caffeine
1	29	23
2	29	21
3	28	20

   a Is the distance the ruler drops longer or shorter when his reactions are faster? [1]

   b Calculate the average distance the ruler dropped before and after caffeine. [1]

   c How did the caffeine affect Chenglie's reaction time? [1]

   d Suggest one way Chenglie could improve his investigation to make his results more trustworthy. [1]

   e Predict how drinking alcohol would affect the distance the ruler dropped before it was caught. [1]

4. Ruby investigates the effect caffeine has on her heart. She measures her heart rate every minute for 5 minutes by taking her pulse.

   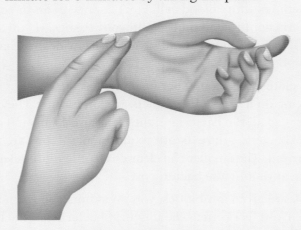

   Then she drinks a can of cola with caffeine in it and continues to take readings every minute. The graph shows her results.

   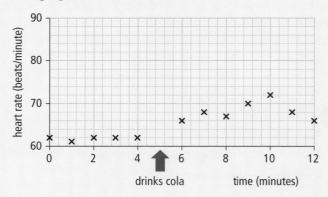

   a Why did she take some readings before she drank the cola? [1]

   b Ruby thinks caffeine affects her heart rate. Is there any evidence for this in her results? [1]

   c Ruby's friend said she didn't have enough evidence. There are other things in cola besides caffeine. One of the other chemicals could be affecting her heart rate. What extra test could Ruby do to make her evidence stronger? [1]

5 The graph below shows how the concentration of alcohol in a driver's blood affects the risk of having a car accident.

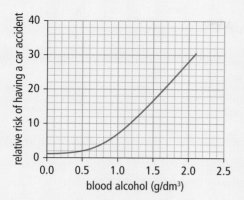

a What is the relative risk of having an accident when the blood alcohol concentration is 1 g/dm³? [1]

b If the blood alcohol concentration doubles to 2 g/dm³, what is the new relative risk? [1]

c A drinker's blood alcohol concentration changes after he drinks alcohol.

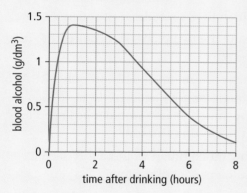

Describe the pattern in the results. [3]

d In India drivers may not have more than 0.3 g/dm³ of alcohol in their blood. How long did the drinker need to wait before he could drive? [1]

6 Scientists compared babies whose mothers drank a lot of alcohol or smoked a lot.

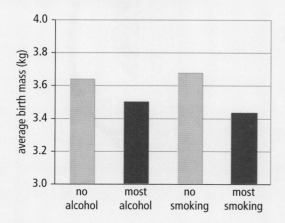

a Which variable did the scientists measure? [1]

b Which had most effect on this variable – drinking alcohol or smoking? [1]

Other scientists compared children whose mothers who drank alcohol when pregnant with children whose mothers did not. The tests were done 1 year after they started school.

Test	Test scores after 1 year of school	
	No alcohol	Alcohol
Maths	96	85
Reading	102	92

c What can they conclude from their results about drinking alcohol when pregnant? [1]

7 Scientists are trying to eliminate these infectious diseases.

Disease	Polio	Dengue	Malaria
Cases per year	1000+ infections	100 million infections	500 million infections
Caught from	infected humans	mosquito bites	mosquito bites
Location	few countries	few countries	many countries

a Which disease infects most people? [1]

b In 2000 the World Health Organization (WHO) recorded the age of each malaria victim who died.

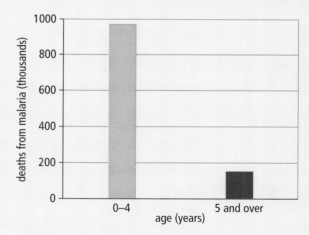

What can you conclude from the results? [1]

c Which of the diseases in the table should to be easiest to eliminate? Explain why. [2]

d Malaria carriers have the pathogen in their blood, but no symptoms. Explain why this makes the disease harder to control. [1]

Review Stage 8

1. Ines wants to prove that plants cannot carry out photosynthesis without carbon dioxide. She sets up the apparatus in the diagram.

   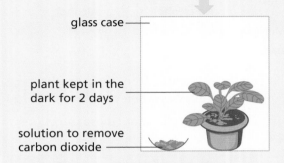

   a Draw the apparatus she should use as a control. [1]
   b What test should she do on each plant's leaves to see whether photosynthesis takes place? [1]
   c Predict what results she will see. [2]

2. The list below shows four parts of a plant.
   A xylem
   B leaf
   C root hairs
   D stomata

   a Write the letters in order to show the route water takes through a plant if it is not used for photosynthesis. [2]
   b Name the substances that plants take in from the water around soil particles. [1]
   c How can a plant reduce water loss when the soil is very dry? [1]

3. The list below shows five nutrients.
   A proteins
   B fats
   C carbohydrates
   D vitamins
   E minerals

   a Select the two main nutrients used as energy sources. [2]
   b Select the two main nutrients used for growth and repair. [2]
   c Which three nutrients cause deficiency diseases when they are not included in a child's diet? [3]
   d This table shows the main sources of vitamin C in a student's diet.

Food	Percentage of vitamin C intake
fruit	30
vegetables	50
other foods	

   Calculate the percentage of their vitamin C that came from other foods. [1]

4. The table shows the main nutrients in three different foods. The rest of the food consists of water, fibre, vitamins, and minerals.

Food	Nutrients present (g/100 g of food)		
	Carbohydrate	Fat	Protein
bread	42	3	9
cheese	0	34	23
banana	23	0	1

   a Which food provides most energy? [1]
   b Name the two types of carbohydrate in a ripe banana. [2]
   c The recommended daily amount (RDA) of protein for girls aged 14–18 is 46 g. How much cheese would a girl need to eat to get this amount? [1]
   d How much bread would a girl need to eat to get 46 g of protein? [1]

5. The diagram shows the human circulatory system.

   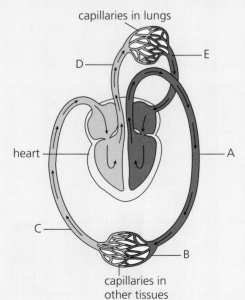

a Which two blood vessels are arteries? [2]

b Which two blood vessels carry blood that is short of oxygen? [2]

c Blood plasma transports urea from the liver to the kidney. Name one other substance carried in plasma. [1]

d Explain why the walls of blood vessel B need to be thinner than the walls of blood vessels A and C. [2]

6 As people age, plaque builds up along the inner walls of arteries. What effect does this have on their blood pressure? [1]

7 Mandeep set up this apparatus to investigate digestion.

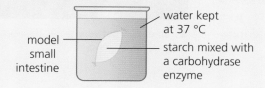

a Why do humans need to digest large molecules like starch? [1]

b Where does starch digestion begin in the human body? [1]

c What does the water represent in Mandeep's model? [1]

d Carbohydrase enzymes break down carbohydrates. What would you expect to find in the water at the end of the experiment? [1]

e Suggest why the water was kept at 37 °C. [1]

8 The list below shows three substances found in tobacco smoke.

A tar

B carbon monoxide

C nicotine

a Which substance destroys cilia and increases smokers' risk of getting lung cancer? [1]

b Which substance combines with haemoglobin in the blood? [1]

c Which substance makes cigarettes addictive? [1]

d Which substance increases the risk that a baby will have a low birth mass? [1]

e One of the substances in cigarette smoke narrows the arteries that take blood to heart muscle.

Drugs and disease

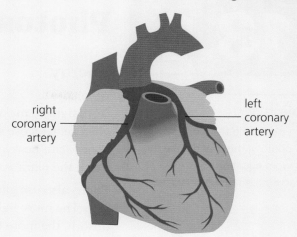

A narrow artery can easily be blocked by a blood clot. Explain why heart muscle cells would die if their blood supply was cut off. [2]

9 The list below shows four parts of the female reproductive system.

A oviduct

B ovary

C uterus

D vagina

a Which letter shows where egg cells develop? [1]

b Which shows where eggs are most likely to be fertilised? [1]

c Which shows where an embryo implants and forms a placenta? [1]

d The graph shows how the lining of the uterus changes when last month's egg is not fertilised.

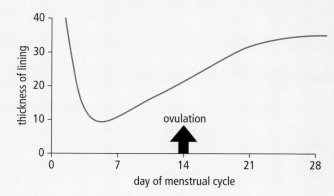

Explain why the thickness decreased between days 1 and 4. [1]

e On which day is the egg most likely to be fertilised? [1]

13.1 Photosynthesis

Objective
- Understand the process of photosynthesis and write the word equation

Energy

All plants need light. They use energy from light to make sugars such as glucose. Glucose stores energy. This energy can be released using respiration when the plant's cells need energy.

Plants also use glucose to build new cells. The more light energy plants absorb, the more they photosynthesise and the more their biomass increases.

Plant biomass keeps every animal alive – including us. It supplies the energy and building materials that herbivores use for growth. Then it gets passed along food chains to carnivores.

↑ Trees can get very tall as they grow towards the light.

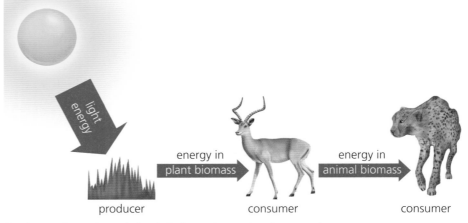

↑ Energy flows through food chains as it transfers from producers to consumers.

Respiration and photosynthesis are opposite processes.
- Photosynthesis makes glucose. Respiration breaks glucose down.
- Respiration in plants and animals uses oxygen. Photosynthesis returns oxygen to the atmosphere.
- Respiration is an **exothermic** reaction – it releases energy. Photosynthesis is an **endothermic** reaction because it takes in light energy.

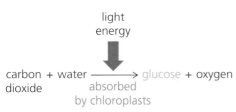

↑ Light energy is absorbed during photosynthesis.

carbon dioxide + water $\xrightarrow[\text{absorbed by chloroplasts}]{\text{light energy}}$ glucose + oxygen

Chloroplasts

Photosynthesis takes place in **chloroplasts**, mainly in leaf cells. These cells take carbon dioxide from the air. Most leaves are very thin so gases diffuse in and out quickly.

The water needed for photosynthesis is brought up from the roots in **xylem** vessels.

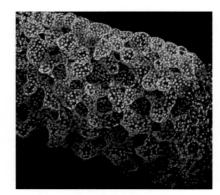

↑ A beam of light made the chloroplasts in these leaf cells glow.

Plants

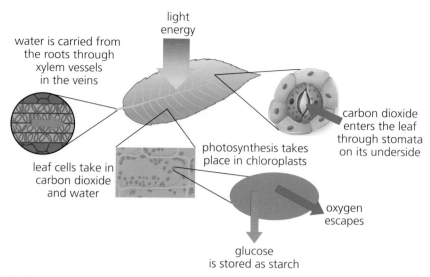

↑ Leaves are very thin so gases diffuse in and out of cells quickly.

Evidence

Seedlings need light. If seedlings are kept in the dark they produce tall, weak shoots and tiny yellow leaves. This growth uses up their energy supply, so the plants die after a few days. Seedlings that receive light have short sturdy stems. They use their first leaves to photosynthesise and produce the glucose they need to keep growing.

Some of the sugars made by leaf cells are moved to cells that have no chloroplasts. The sugars travel through the plant in **phloem** tubes. Other sugars are converted to starch and stored in leaves. **Iodine** can show where photosynthesis has taken place. It turns any stored starch dark blue.

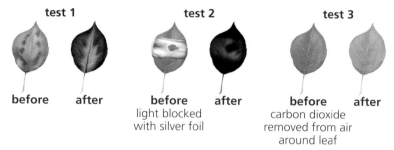

↑ If leaf cells contain starch they must have carried out photosynthesis.

↑ Seedlings grown in sunlight.

↑ Seedlings grown in the dark.

Q

1. Explain why animals depend on plants to keep them alive.
2. Describe how carbon dioxide and water reach the chloroplasts in leaves.
3. Describe two pieces of evidence that show that plants cannot make their own food without light.
4. Starch tests show where photosynthesis has taken place. What can you conclude using the evidence from test 1?
5. What does the evidence from test 3 show?
6. The amount of oxygen in a sample of air can be measured using an electronic sensor. How would you expect it to change around a sugar cane plant on a sunny day?

!

- The equation for photosynthesis is:
  carbon dioxide + water
  → glucose + oxygen
- Leaf cells absorb the carbon dioxide they need from air.
- Water is transported from the roots in xylem vessels.
- Glucose stores energy.
- Stored energy is transferred along food chains to animals.

Enquiry 13.2

Preliminary tests

Objectives

- Suggest and use preliminary work to decide how to carry out an investigation
- Decide which measurements and observations are necessary and what equipment to use

Deciding what to investigate

Salma has a question about photosynthesis. She knows that light is essential, but does the colour of the light matter?

She finds some information about light: 'White light is made up of red, green, and blue'. This gives her an idea:

Leaves look green, so they must absorb the red light and blue light.

If they absorbed every colour there would be no light left, so they would look black.

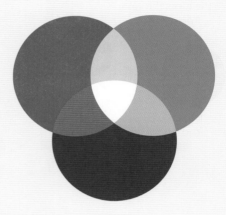

RGB

⬆ White light is made up of red, green, and blue.

Choosing apparatus

Salma collects a plant, a powerful lamp, and some filters. Each filter lets a different colour through.

Now she needs a way to measure photosynthesis. She would like to do all her tests on the same leaves on the same day. That will make it a fair test.

Salma does some research on the internet. She finds a way to measure photosynthesis very quickly. It uses tiny leaf discs. The discs normally float because leaves are full of air (see page 84). If you suck the air out using a syringe, they sink.

If the leaf discs are photosynthesising, they make oxygen. Bubbles of oxygen collect inside the discs and make them float again. The faster the leaf discs rise, the higher the rate of photosynthesis.

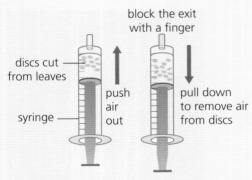

⬆ A syringe can be used to suck the air out of leaf discs.

Preliminary tests

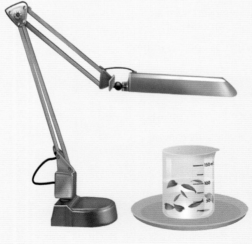

filters can be placed over the beaker to change the colour of the light

➡ The higher the rate of photosynthesis, the less time the discs take to rise.

Salma draws a diagram to show how she will set up the apparatus. She knows what she will change and what she will measure. She spots two important variables she needs to control – the volume of water and the distance between the lamp and the discs.

160

Salma does some preliminary tests using white light.

The volume of water does not affect the results but the distance from the lamp makes a big difference.

Distance from lamp (cm)	Time taken for a disc to rise (minutes)
10	3
20	12
30	29

Salma decides to use 50 cm³ for every test and put the lamp 10 cm from the discs. That will make them rise quickly, but not too fast to measure.

She decides to use 10 discs with each colour. That will make her results more reliable. She will need to calculate the average time they take to float.

Salma will check how many discs are floating after every minute. Then she will record the time when half the discs are floating.

She predicts that red light and blue light will make the discs rise faster than green light.

Evaluating the results

Salma's results agree with her prediction.

Filter used	Time taken for discs to rise (minutes)
blue	5
green	35
red	9

To make her evidence stronger she leaves some discs in the dark. After 3 hours they are still at the bottom of the beaker. This shows that the discs only rise when they receive light.

Salma was surprised that red and blue filters gave different results. Perhaps there was another variable she did not control. She realises that filters absorb some of the light from the lamp. She should have used a light meter to check that each coloured light had the same intensity.

Q

1. What question about photosynthesis does Salma plan to answer?
2. Why does Salma think that green light is no good for photosynthesis?
3. Draw a diagram of the apparatus Salma sets up.
4. Explain how Salma's method works.
5. Salma carried out preliminary tests before she started her investigation. Why did she do this?
6. Salma expected red and blue light to give similar results. Suggest two possible reasons why they gave different results.

- In any investigation, some variables need to be controlled.
- Preliminary tests check that the method works and the controlled variables have suitable values.

13.3 Plant growth

Objective
- Understand the importance of water and minerals to plant growth

Water

These strawberry plants are growing in water-filled pipes high above the ground. As their leaves lose water by **transpiration** (page 88), more water moves up through xylem vessels from their roots to replace it. The roots take in fresh water from the pipes.

Growing plants like this is called **hydroponics**.

⬆ The roots of these plants take water from the pipes they grow in.

Minerals

When plants take in water from soil, they also take in essential minerals. These minerals come from dissolved rocks, decaying plants and animals, or artificial fertilisers. Different soils contain different amounts of each mineral. For example, sandy desert soils do not usually contain much nitrogen.

In hydroponic systems, growers can add just the right amount of each mineral to the water.

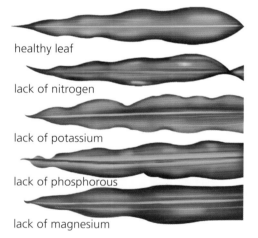

⬆ The top maize leaf was grown in fertile soil. The others were grown in soil lacking one essential mineral.

Essential minerals

To find out which minerals plants need, scientists start with a mixture of minerals and take one away at a time. Each mineral contains a different element. Together the elements provide everything a plant needs for healthy growth.

A control group is given a complete mixture of minerals. Test groups are given mixtures with different minerals missing.

Mineral element	How the plant uses it	Deficiency symptoms
nitrogen (in nitrates)	used to make proteins to build new cells	older leaves turn yellow.
potassium	used during photosynthesis and respiration	poor flower and fruit growth. Older leaves wilt and begin to lose colour.
phosphorus (in phosphates)	involved in energy transfer during photosynthesis and respiration	poor root growth. Leaf tips turn brown and older leaves turn purple.
magnesium	used to make chlorophyll	older leaves turn yellow between the veins.

Evidence from many investigations like this confirms that the most important minerals are the ones that contain nitrogen, potassium, phosphorus, and magnesium.

Aeroponics

↑ The roots of these rice plants hang in air so they can absorb oxygen.

Roots take in minerals by **active transport**. The minerals can't simply diffuse into plant cells like oxygen does. They need to be moved from where they are very spread out (in soil) to where they are concentrated (in plant cells). So the roots need to use energy when they take in minerals.

To release the energy they need for active transport, roots must respire – so they need oxygen. When roots grow in soil they take oxygen from the air between soil particles. They can take some oxygen from water, but oxygen isn't very soluble.

Aeroponics lets roots get extra oxygen. The roots are left dangling in air. They are sprayed with water and dissolved minerals every few minutes. The roots stay moist and get as much oxygen as they can use.

Aeroponics makes plants grow faster and use less water.

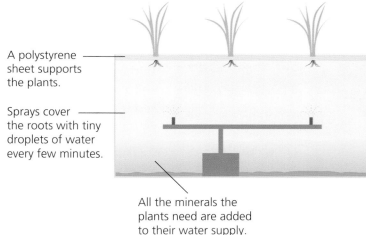

A polystyrene sheet supports the plants.

Sprays cover the roots with tiny droplets of water every few minutes.

All the minerals the plants need are added to their water supply.

↑ Plants obtain all their water and minerals from a spray.

Q
1. What is added to the water used in hydroponic systems?
2. List the four main elements plants obtain from minerals.
3. A plant has purple leaves and few roots. Which mineral is it short of?
4. List the symptoms that show a plant is short of potassium.
5. Plants grown in soil lacking nitrogen are usually very small. Explain why they can't grow properly without nitrogen.
6. Describe how you could collect evidence to show that rice plants need magnesium.

- Roots replace the water plants lose by transpiration.
- Roots also take in minerals containing nitrogen, potassium, phosphorus, and magnesium.

Extension 13.4

Phytoextraction

Objectives
- Understand the importance of water and mineral salts
- Explain results using scientific knowledge and understanding

Cleaning up with plants

The Chernobyl nuclear power plant exploded in 1986. Clouds of radioactive particles escaped. The highest radiation levels were in lakes and forests close to the accident. They were contaminated with a radioactive metal called uranium. Large amounts of radioactivity can kill or cause serious health problems.

↑ When this nuclear power plant exploded it scattered radioactive uranium across the surrounding countryside.

Scientist turned to plants to clean up the water. They floated rafts of sunflowers on top of the lakes and waited. The plants pulled radioactive uranium into their roots. When they couldn't take in any more, they were taken to a safe place and buried.

Using plants to remove metals in this way is called **phytoextraction**.

Hyperaccumulators

Plants have a natural ability to absorb minerals. They need the minerals to build their cells. The most useful ones are nitrates (containing nitrogen), phosphates (containing phosphorus), potassium, and magnesium. Plants also need minute amounts of other metals, but not too much. Many metal compounds damage living things.

In the past, the process of extracting metal from the earth in mines polluted soil and water. There are still many places where nothing will grow as a result of mining activities. The soil contains compounds of copper, zinc, lead, cadmium, or arsenic.

↑ Nothing grows here because the soil contains high concentrations of lead, copper, tin, and zinc.

A few plants can survive in polluted soils. They absorb a lot of the metal, but it doesn't harm them. Plants like these are called **hyperaccumulators**.

metals spread through a large volume of soil → bioaccumulation in the plant's leaves

↑ Hyperaccumulators extract metals from large volumes of soil.

Plants

Storing metals

When a cell has enough water, its vacuole presses out against the cell wall. This keeps the cell firm. If a plant's cells are firm like this, the plant can support itself.

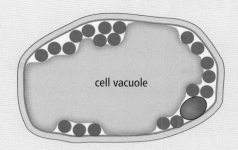

← Plant cells keep their vacuoles full of water so the cells can support themselves.

Water evaporates from leaves and their cells use water for photosynthesis. Xylem vessels bring fresh water and minerals up from the roots. The minerals are stored in leaf cell vacuoles until they are needed. If the plant is growing on soil containing toxic metals, these metals accumulate in the vacuoles of leaf cells too. If the plant is burned, the metals are left in the ash.

If we grow hyperaccumulators on contaminated soil, the metals that were spread though a huge amount of soil are left in a tiny quantity of ash. Any valuable metals can then be extracted from this ash.

Improving health

In many parts of Bangladesh, the water is contaminated with arsenic, which is a very poisonous element. Over the years it can accumulate in people's bodies. It causes serious health problems.

These Chinese ferns accumulate arsenic. The concentration in their leaves can be 1000 times higher than in the soil.

Now scientists want to plant millions of the ferns along riverbanks in Bangladesh. They could make the water safer by extracting arsenic from it.

Arsenic is needed to make solar panels, so the scientists plan to extract the arsenic from the ferns and sell it.

↑ This Chinese fern accumulates arsenic in its leaves.

1 Give two reasons why some metals need to be removed from soil or water.
2 Explain why some plants are called hyperaccumulators.
3 Describe how hyperaccumulators can be used to remove metals from soil.
4 Explain how ferns that accumulate arsenic could improve the health of villagers in Bangladesh.

- Plants need minerals containing nitrates, phosphates, potassium, magnesium, and minute amounts of other metals.
- Minerals are stored in cell vacuoles until they are needed.
- Toxic metals can also be taken in and stored by plants.
- Hyperaccumulators store more metals than other plants.

13.5 Flowers

Objectives
- Recognise each part of a flower and describe its function
- Understand how pollination and fertilisation take place
- Distinguish between insect-pollinated and wind-pollinated flowers

Reproductive organs

Flowers contain male and female sex organs. They let plants reproduce by making seeds.

Seed development begins when a male sex cell fertilises a female sex cell. Some plants produce male and female sex cells in separate flowers. Others have their male and female sex organs in the same flowers.

↑ This flower contains both male and female organs.

The male parts of a flower

The male parts of a flower are called **stamens**. The most important part of a stamen is its **anther**. This produces **pollen** which carries the male sex cell. Each anther has a **filament** to hold it in a good position to spread its pollen.

The female parts of a flower

The female part of a flower is called a **carpel**. At the base of each carpel there is an **ovary**. This produces **ovules** which contain female sex cells. Above the ovary is a sticky **stigma** which catches pollen. The **style** separates the stigma from the ovaries.

Pollination

Male and female sex cells are called **gametes**. To bring them together, pollen needs to be moved from a stamen to a stigma. This is **pollination**.

Many flowers can self-pollinate. This happens when pollen moves to a stigma in the same flower, or a stigma in another flower on the same plant. For the long-term future of the species, cross-pollination is better. This means the pollen moves to the stigma of a flower on another plant. It gives the plants' offspring a greater variety of characteristics.

↑ Self-pollination occurs when pollen land on the stigma in the same flower or on the same plant.

To stop self-pollination happening, flowers can produce their male and female gametes at different times.

Using insects

Many flowers use insects such as bees to spread their pollen. They use scents, coloured petals, and sugary nectar to attract the bees.

When a bee visits a flower it gets covered with sticky pollen grains. As it moves from flower to flower, it leaves this pollen on their stigmas.

↑ Bees transfer pollen as they collect nectar from each flower.

Plants

Relying on the wind

Most grasses and many trees rely on the wind to spread their pollen. They do not need to attract insects so their flowers lack petals and nectaries. Their pollen is small and light and their stamens and feathery stigmas hang out to catch the wind.

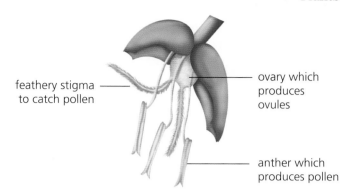

↑ Grass flowers rely on the wind to carry pollen from flower to flower.

Fertilisation

A pollen grain begins to grow when it lands on a stigma. It sends a pollen tube down through the style to an ovule. The male gamete's nucleus moves down through this tube, enters an egg cell, and fuses with the egg cell nucleus. This is **fertilisation**.

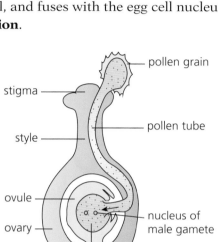

← A pollen tube grows down through the style to carry the nucleus of the male gamete to the egg cell.

The fertilised egg grows into an embryo, and the ovule produces a **seed** to protect it. As the seed develops, the ovary forms a **fruit** around it. If the ovary wall becomes very hard, the fruit is called a nut.

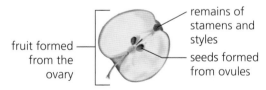

↑ After fertilisation, seeds form and a fruit develops.

- The male parts of flowers are stamens.
- Each stamen has pollen-producing anther on a stalk called a filament.
- The female part of a flower is a carpel.
- A carpel has pollen-collecting surface called a stigma.
- A style connects the stigma to an ovary which produces ovules.
- Insect-pollinated flowers have colourful petals and nectarines to attract insects.

Q

1. Draw diagrams to show the male and female sex organs for each type of flower:
   a) a flower pollinated by insects
   b) a flower pollinated by the wind.
2. List the main differences between insect-pollinated and wind-pollinated flowers.
3. What is different about the pollen of insect-pollinated and wind-pollinated flowers?
4. Describe how flowers can make self-pollination difficult.
5. Describe the differences between pollination and fertilisation.

167

13.6 Seed dispersal

Objective
- Understand seed dispersal in flowering plants

Germination

Cool, dry seeds can last for years. The embryos inside them are **dormant**. That means they are alive but not growing. They start to grow when they have water, a suitable temperature, and oxygen. This is **germination**. Some seeds also need light to germinate.

↑ A seed germinates when the embryo inside it starts to grow.

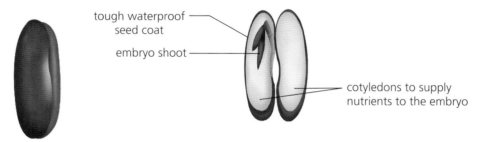

↑ Each seed has a store of nutrients.

While it is dormant an embryo inside a seed can travel long distances and survive conditions that would kill a growing plant, such as extreme temperatures or lack of water.

Seed dispersal

Seedlings would struggle to grow in the space between these tree trunks. There is too much competition for light and water. Older trees block off most of the light and extract nearly all the water from the soil. Any seedlings that germinated here would not survive for long. To increase their chance of survival, they need to get a long way from their parents.

Spreading seeds away from the parent plants is called **seed dispersal**. Some plants rely on the wind, or water, to carry the seeds away. Some use animals. Others simply shoot their seeds into the air.

↑ Seedlings struggle to survive in the dry soil and dim light between these trees.

Exploding pods, wind, and water

Pea seeds are produced in dry fruits called pods. When the seeds are ripe, the pods burst and peas scatter in every direction.

Coconut palms and mangrove trees grow near water. Their fruits float. When they roll into the sea they can be carried across oceans.

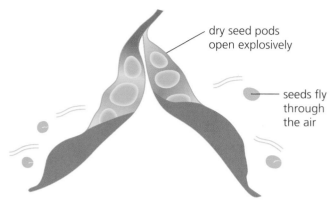

⬆ When pods explode, seeds can be thrown long distances.

Seeds dispersed by the wind are small and light. Some have wings. Others have feathery parachutes. They are usually produced in very large numbers.

⬆ Wings and parachutes keep seeds in the air longer so they can travel further.

Using animals to spread seeds

Some fruits stick to animals' feet. Others have hooks. They stick to animals' fur. Fruits like these can be carried a long way before they rub off.

Many insects and small animals collect seeds and nuts. They hide them for later. The ones they forget about can germinate in their hiding places.

Other seeds attract animals by producing sweet juicy fruits. When the animals eat the fruits, any large seeds are thrown away. Smaller seeds can pass through animals' digestive systems without being damaged. They are carried wherever the animal goes and deposited in their faeces.

Most of the plants in tropical forests use animals to spread their seeds – mainly birds, bats and monkeys.

⬆ Hooked fruits are carried accidently but nuts are deliberately collected.

Q
1. What conditions are needed to make a seed germinate?
2. Explain why seedlings have a better chance of survival if they grow a long way from their parent plant.
3. Sketch fruits that could be carried by the wind, water, and animals. Label any useful characteristics they have.

- Seed dispersal helps to reduce competition between plants of the same species.
- Seeds can be spread by exploding seed pods, wind, water, or animals.
- Some animals deliberately collect seeds and nuts.
- Others carry sticky fruits and seeds accidentally.
- Many plants rely on their fruits being eaten and their seeds deposited elsewhere.

Review 13.7

1. The diagram shows the things that leaves absorb, and the things they make, when their cells photosynthesise faster than they respire. Identify A–E. [5]

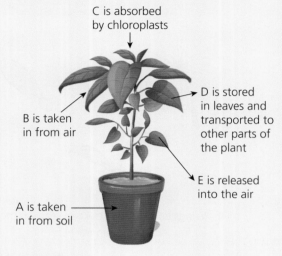

2. Choose the correct statements about photosynthesis: [2]
   a. Photosynthesis takes place in chloroplasts.
   b. Photosynthesis is carried out by consumers.
   c. Photosynthesis makes glucose.
   d. Photosynthesis releases energy.
   e. In plants, photosynthesis is always faster than respiration.

3. Stephan grew seeds in glass boxes containing different amounts of carbon dioxide.

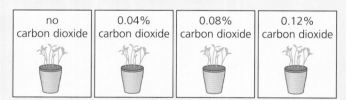

He dried the plants in each box and measured their mass. Then he subtracted the mass of the original seeds to see how much new biomass they made.

Carbon dioxide (% in air)	Plant biomass produced (g)
0.00	0.00
0.04	0.09
0.08	0.18
0.12	0.21

Stephan plotted a graph of the results.

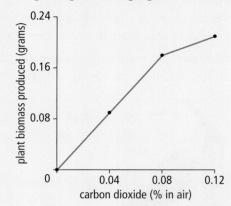

a. What conclusion can you make from Stephan's graph? [1]
b. Write the equation for photosynthesis. [2]
c. Use your knowledge about the equation for photosynthesis to explain the results. [1]
d. Suggest one thing Stephan could do to make his results more reliable. [1]

4. Mariam wants to know which minerals are important for plant growth. She takes four samples of pure water and adds different minerals to each of them. Then she puts a seedling in each sample of water.

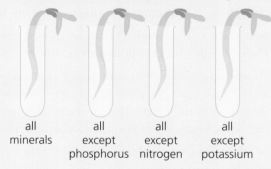

After 2 weeks she compares the masses of the seedlings.

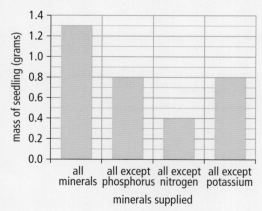

a. Suggest one thing Mariam kept the same for every mineral solution she tested. [1]

b Suggest one other thing she could have measured apart from the total mass of the seedlings. [1]

c Which mineral appears to be most important for growth? [1]

d What is this mineral used for? [1]

e The green substance in chlorophyll contains magnesium. Suggest how the colour of leaves would change in plants grown without magnesium. [1]

5 The diagram shows the main parts of an insect-pollinated flower. Choose the correct letter for each of these parts.

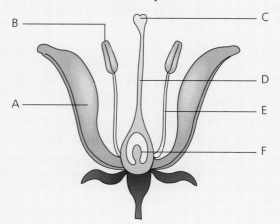

a anther [1]
b filament [1]
c ovary [1]
d stigma [1]
e style [1]
f petal [1]

6 The diagram shows the main parts of a wind-pollinated flower.

Describe two differences between wind-pollinated flowers and insect-pollinated flowers. [2]

7 Joshua is trying to find the concentration of sugar needed to make pollen grains grow. He makes different sugar solutions. He leaves 10 pollen grains in each solution for 4 hours. Then he observes them under the microscope and records how many of them grow pollen tubes.

Sugar solution concentration (g/dm³)	Number of pollen tubes that grew
0	0
50	2
100	8
150	1

a Which variable did Joshua change? [1]

b What percentage of the pollen grains grew a pollen tube in the 100 g/dm³ sugar solution? [1]

c Joshua thought the result for the 150 g/dm³ sugar solution might be anomalous. What should he do to make his results more reliable? [1]

8 Salma is investigating methods of seed dispersal. She has six types of seed.

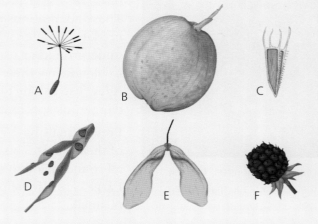

a Choose two seeds that are dispersed by the wind. [2]

b Which seed is dispersed by water? [1]

c Which two seeds are dispersed by animals? [2]

d Which seeds are shot into the air when their seed pod bursts? [1]

14.1 Adaptation

Objective
- Explain the ways in which living things are adapted to their habitats

Naked mole rats

Naked mole rats look rather odd, but they are well adapted to their habitat. They live in large underground colonies under the desert. They rely on smell, touch, and hearing to find their way in the dark.

Mole rats let their body temperature vary. This saves energy. They have a low respiration rate and need very little food and oxygen. They huddle together to stay warm and move deeper underground if they get too hot.

↑ Naked mole rats' teeth stick out so they can dig tunnels without getting soil in their mouths.

Only the queen mole rat breeds. The rest of the mole rats search for food or guard their tunnels. They get all their water from the roots they eat.

Like the mole rats, all animals need to find food and avoid predators. The specialised characteristics that help them do this are **adaptations**. They can be physical features or behaviours. Most adaptations take many generations to develop.

Predator or prey?

Prey animals have adaptations that make them hard to see, hard to catch, or difficult to eat. They might have spines, hard shells, nasty tastes, or produce toxins. These adaptations put off most predators. Defenceless prey can hide underground or use **camouflage** to blend with their surroundings. A few use colour to mimic more dangerous species.

↑ Predators and prey have different adaptations.

↑ Leafy seadragons are fish but their camouflage makes them look like seaweed.

Prey animals have eyes on the sides of their heads. This gives them good all-round vision to spot predators. Large ears also help, or a good sense of smell.

Living in large groups makes life easier for prey. They can take turns to look out for predators, which gives them more time for eating. A predator will only catch one group member – the slowest or weakest.

↑ Young zebras are ready to run within minutes of birth.

Predator animals have different adaptations. Forward-facing eyes help them locate their prey. Weapons are also useful, like a sharp beak, teeth, or claws. Predators can also use camouflage to avoid detection.

Moving around

In open spaces, speed is a useful adaptation. It can help predators catch their prey, and help prey to avoid being eaten. Fast land animals usually have long legs and efficient lungs. Those that live in water have strong tail muscles.

The fastest animals have smooth **streamlined** shapes that are narrow at both ends.

In crowded forests speed is not so useful. Animals that live in trees need a good sense of balance and feet that can grip the branches.

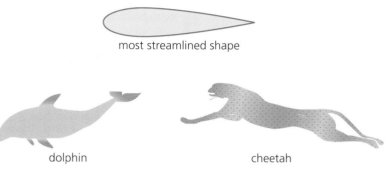

⬆ A streamlined shape helps animals move faster.

webbed feet help ducks swim

wide feet stop bears sinking

ape feet grip branches

soft hooves give mountain goats a good grip

⬆ Animals have very different feet. They are adapted for where the animal lives.

On loose sand or snow, wide feet are useful. They don't sink in. Animals that push themselves through water, or dig, also need wide feet. On ice or steep mountainsides a good grip is essential.

Animals that fly or glide need to be lightweight and have a large surface area.

Q
1. Explain what an adaptation is.
2. Suggest adaptations that would help these animals survive:
   a. a predator that lives in the sea
   b. a flying predator
   c. a prey animal that lives in the treetops
   d. prey animals that live on open grassland
   e. a predator that hunts on open grassland.

- Adaptations are specialised characteristics that help an organism survive.
- Adaptations help animals to find food and avoid predators.

14.2 Extreme adaptations

Objective
- Explain the ways in which living things are adapted to challenging environments

Life in the Sahara desert

Living things struggle to survive in the Sahara. It is one of the driest placest on Earth – hot during the day but cold at night. To survive there, plants and animals need to find water and hold onto it.

Some desert species have very short lives. They grow and reproduce quickly when it rains. It is easier for seeds or eggs to survive for long periods without water. Some lie dormant for many years before it rains again.

↑ This desert plant can survive for long periods without water. Other desert plants store water in swollen stems or fleshy leaves.

Saving water

Most desert plants have long roots near the surface. They absorb rainwater quickly before it soaks into the sand. They also have tiny leaves, spines, or a waxy coat to stop water evaporating from their leaves.

Some desert plants use swollen stems or fleshy leaves to store water. Others dry out but can recover when it rains.

Most desert animals don't drink. They get water from their food or from stores of fat. They stay underground during the day to keep cool. They don't sweat and their urine is very concentrated.

↑ Desert animals like this lizard extract the water they need from their food.

Surviving in the Arctic

Arctic temperatures are very low, especially in winter when it stays dark for months. Even in summer the soil is frozen below the surface.

The only plants that survive here are mosses and small clumps of flowering plants. These store nutrients to keep them alive through the winter. They grow clustered together and use hairs or dead leaves to protect new shoots from the icy wind.

Where plants can't survive there are tough, slow-growing lichens. Each lichen is a fungus with algae in its tissues, so lichens produce their own food like plants do.

Animals need to keep warm to survive in the Arctic.

↑ In winter reindeer scrape away the snow to find food.

Adaptation and survival

Reducing heat loss

Birds and mammals keep their bodies at a set temperature. A large rounded body, fat, and thick feathers or fur help them to retain heat.

A large body holds more heat and a rounded shape cools down more slowly. Long thin bodies with large ears lose heat quickly.

Layers of fat, and the air trapped in fur, are both good **insulators**, so they also cut heat loss. Small animals can burrow under snow to get extra insulation. The snow traps air around them and protects them from icy winds.

Animals also use behaviours to control their temperature. They can keep warm by huddling together.

↑ Lemmings spend winter under the snow to avoid icy winds.

Life on the ocean floor

Deep in the oceans it is cold and dark and there isn't much oxygen. Dead animals falling from the surface are the main source of food.

Many of the animals here use feelers or smell instead of eyes. They move slowly to save energy and can survive for months without food.

Some fish use chemical reactions to create light so they can see. Others use light to attract prey or find mates. The light is produced by bacteria living inside their tissues.

↑ Anglerfish use light to attract prey.

Q
1. Draw a desert plant and label the adaptations that help it survive.
2. Some small desert animals can survive for a long time without drinking. List three adaptations that help them to do this.
3. Describe three adaptations that help arctic animals conserve heat.
4. In milder winters, when there is less snow, fewer lemmings survive. Suggest why.

- Adaptations that save water help organisms survive in hot dry places.
- Some desert plants have long horizontal roots, thick stems, small fleshy leaves, spines, or waxy coatings. Others are able to dry out without dying.
- Desert animals extract water from food, avoid sweating, and produce concentrated urine.
- Adaptations that conserve heat help animals survive in cold places.
- Arctic animals have rounded bodies and thick layers of fat and fur.

Extension 14.3

Survival

Objective
- Recognise that existing adaptations can become less useful if the environment changes

Climate change

Quiver trees are adapted to desert life. They have long shallow roots and their stems and leaves store water. But gobal warming is raising the temperature in their habitat and reducing the rainfall. As their environment gets hotter and drier, quiver trees struggle to survive. Many of the trees nearest the equator are dying.

↑ As their environment gets hotter and drier, these trees struggle to survive.

As the climate warms, many species move further from the equator, or up mountainsides, to cooler habitats that suit their adaptations. Plant movements take many generations because growing plants can't move. Species move gradually as more of their seeds survive in cooler places.

Adaptations can become less useful when the environment changes. If a species can't move or adapt, it becomes extinct.

Changing crops

Future climate change could make it harder for us to grow the plants we need for food.

The world's three main crops are maize, wheat, and rice. In future more countries may need to grow crops adapted to drier conditions, such as bananas, cassava, and cowpeas.

Changing the environment

Temperatures are rising fastest at the poles so the environment there is changing rapidly.

Polar bears live in the Arctic. Their adaptations help them to stay warm and make them successful predators. Their main prey are seals.

Seals hunt under sea ice and they are fast swimmers. Polar bears can't catch them in the water. Instead they lie in wait near holes in the ice. When a seal comes up to breathe, the bear tries to catch it.

↑ Global warming makes it harder for polar bears to catch their prey.

The ice that bears hunt on melts in the summer, so they go without food for months. As the Arctic gets warmer, this ice melts earlier and the bears have less time to hunt. They get thinner and raise fewer cubs. The adaptations that made them successful are no use when there is no ice.

Most animals can survive small temperature rises, but climate change can make it harder for them to find food.

Adaptation and survival

Coping with change

Changes in the environment often affect one species more than another.

Adelie and chinstrap penguins nest in western Antarctica. Both species feed on krill but they have different physical adaptations and different behaviours:

Adelie penguins breed in the summer when the snow has melted. They are slow swimmers but they can hold their breath for a long time. They feed under sea ice where there are fewer predators.

↑ Adelie penguin.

↑ Chinstrap penguin.

Chinstrap penguins breed in spring. They can swim fast but they can't hold their breath for very long. They feed in open water.

As the Antarctic gets warmer, sea ice forms later and melts sooner, and more snow falls. The graphs show how these changes affect each species of penguin.

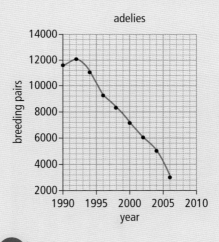
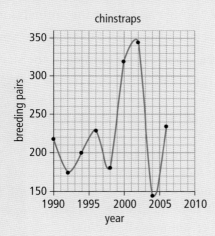

← The graphs show the number of breeding pairs in western Antarctica.

Q

1. Many plants are growing in areas further towards the poles as the climate warms. Explain why.
2. Explain why climate change could affect our food supply.
3. Explain how global warming has reduced the number of seals that polar bears catch.
4. Adelie and chinstrap penguins both breed in Antarctica. How did their numbers change between between 1990 and 2000?
5. Which penguin species finds it harder to catch its food when there is less sea ice?
6. Adelies breed later than they used to. Explain why.
7. Adelie chicks grow less than they used to before winter comes again. Explain why this reduces their chance of survival.

- When the environment changes, existing adaptations can become less useful.
- Species that can't move away or adapt become extinct.
- Environmental changes can affect some species more than others.

Enquiry 14.4: Sampling techniques

Objective
- Describe how sampling techniques can be used to estimate populations

Counting animals

Christian's helicopter zig-zags across the park. He is counting elephants. They are large and slow moving so they are easy to count from the air.

Each elephant eats around 150 kg of food a day. It's important to know how many elephants there are in an area. If there are too many they will run out of food and then starve.

If the elephant numbers fall, Christian will try to find out why. That usually only happens when criminals kill them to steal their tusks.

⬆ Elephants can be counted from the air.

Estimating animal numbers

Other animals are harder to count. They may hide, or there may be too many to count. Scientists can estimate population sizes using the **mark and recapture** method. This involves four steps:

1. Capture part of the population, mark them, and release them.
2. Capture another sample of the animals later.
3. Calculate what fraction of the second sample is marked.
4. Divide the number you marked at the start by this fraction.

Dina used this method to estimate the number of meerkats in a colony. She used food to attract them. She caught 12 meerkats. She marked each of the meerkats by fitting a band around one of their legs, and then released them. The next day she caught six meerkats. Two of them were marked, but four were animals she hadn't seen before. She used her results to estimate the total population.

⬆ Two of these meerkats are marked, four are not. So the marked animals represent one-third of the total population.

Number marked = 12

Number caught second time = 6

Fraction that were recaptured = $\frac{2}{6} = \frac{1}{3}$

Estimated population = $12 \div \frac{1}{3} = 12 \times \frac{3}{1} = 36$

Capturing small animals

Axel is studying invertebrates. He makes a pitfall trap to catch any that run along the ground. The trap is just a bucket that animals can fall into. He covers the trap to protect trapped invertebrates from rain, sunlight, and birds.

To collect animals that live in bushes he uses a sweep net. He swings the net through the bushes. Then he empties everything he catches onto a white sheet to make the invertebrates easier to see.

↑ Invertebrates can be sampled using pitfall traps and sweep nets.

Estimating plant numbers

Kristen is investigating plant growth near a tree.

She places a **quadrat** beside the tree. The quadrat is a square frame. It encloses an area of exactly 1 m². The ground underneath the quadrat here is completely bare so she records a percentage plant cover of zero. She keeps moving the quadrat further from the tree, in 5 m steps. In each position she records the percentage of the soil covered with plants. Finally she plots a bar chart to display her results.

Kirsten has an explanation for her results. Trees stop light reaching the ground, so few plants can grow near trees.

Quadrats are also used to estimate plant population sizes. This involves four steps.

1. Measure the total area in m².
2. Place a 1 m² quadrat in a random position.
3. Count the number of plants inside it.
4. Multiply the total area by the number of plants in 1 m².

To make the results more reliable, you should count plants in more than one quadrat. Then use the average number of plants in a quadrat to estimate the total population.

↑ Plants can be sampled using a quadrat like this.

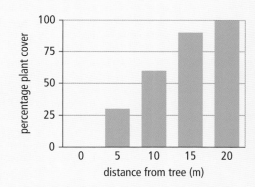

↑ A quadrat can be used to sample the plants in different places.

Q

1. Scientists want to know how many rabbits live on an island. Explain how they could estimate the size of the population.
2. How could you estimate the number of plants on 1000 m² of land?
3. Draw a pitfall trap and explain how it works.
4. Study the results of Kirsten's investigation. What do her results show?
5. What explanation does Kirsten give to explain her results?

- You can sample plants and animals to estimate the sizes of their populations.
- Animals are sampled by trapping them and plants are sampled using quadrats.
- You can estimate animal populations using the mark and recapture method.

14.5 Studying the natural world

Objective
- Understand how scientists study the natural world

Observing animals

Ecologists like Jane Goodall study animals in their natural environment.

Jane studied a troop of chimpanzees in Africa for 34 years. Her findings changed people's ideas about the animals completely. They were not the simple fruit eaters people had imagined from watching zoo animals.

Chimpanzees fought, hunted smaller mammals, made tools, used medicinal plants, and learned from each other.

Each group of chimpanzees is led by a large male. He gets extra food, the best sleeping place, and more chances to mate. The chimps in a troop remember where they fit in socially. Males have to fight for the top position and need to be popular, so they spend a lot of time socialising. They use gestures, calls, facial expressions, and body language to communicate, and they clean each other's fur.

↑ Jane Goodall revealed that chimpanzees had complex social lives.

Two other scientists studied one species for many years. Dian Fossey studied gorillas in Rwanda and Biruté Galdikas studied orang-utans in Borneo. They showed that these animals also had social lives like humans.

Following tracks

Snow leopards live in the mountains of central Asia. They stay hidden so they are rarely seen. Their main prey are wild sheep and goats. There are fewer wild prey animals now because more farm animals are raised in the mountains. Farmers shoot snow leopards because they sometimes kill farm animals for food.

Scientists can estimate snow leopard numbers by following their tracks and collecting their faeces. Laboratory tests on the faeces can show which animal they belong to and what it has been eating.

↑ Snow leopard tracks and faeces show how many there are and what they eat.

Adaptation and survival

Scientists help farmers to protect their farm animals. They also encourage tourists to visit their villages to see the leopards. They want farmers to earn more from tourists so that they can afford to protect the leopards.

Using automatic cameras

There are five species of rhinoceros. Three species are critically endangered, which means there are very few left. In the past, the rhinos were killed for their horns.

The Javan rhinoceros lives in rainforests. There are only a few of them left. Scientists use automatic cameras to video them. The animals don't notice the cameras so they continue to behave naturally. The cameras can be triggered by movement or the animals' body heat.

Scientist hope to learn more about rhino behaviour. It could help them to protect the rhinos. They can also use the cameras to spot poachers.

↑ In the past many rhinos were killed for their horns.

Using electronic tags

Sea turtle numbers are dropping and scientists want to find out why. The turtles can only be observed easily when they return to the shore to breed. They spend most of their lives out in the ocean.

Scientists can study turtle movements by attaching electronic tags to their backs. The tags transmit signals to satellites orbiting the Earth. These show exactly where each turtle is, but the tags are very expensive.

After this turtle was tagged she travelled 5000 km and visited seven different countries. Researchers plan to protect parts of the ocean where most turtles spend their lives.

↑ Sea turtles travel thousands of kilometres each year.

Q

1. List four methods scientists use to study animals in their natural environment.
2. Suggest which method would be best for studying each of the following:
   a. how a lioness brings up her cubs
   b. the migration of wildebeest
   c. shy mammals in a tropical rainforest
   d. whale migration
   e. the number of arctic foxes in a habitat.
3. What can scientists learn from studying animal faeces?
4. What can scientists learn from long-term observations of a group of animals.

Ecologists study animals in their environments by:
- direct observation
- studying their tracks
- using automatic cameras and electronic tags.

Review 14.6

1. These four animals have different adaptations.

   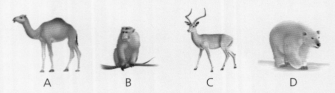

   A   B   C   D

   Match each animal to a habitat it could survive in. Choose from a desert, rainforest, arctic region, or grassland. [4]

2. The diagram shows a hawk.

   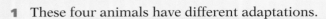

   List three physical adaptations that help make it a successful predator. [3]

3. Fox A lives in the Arctic and fox B lives in the desert.

   A   B

   a. Fox A has brown fur in the summer and white fur in winter. Explain how this colour change helps it survive. [1]

   b. Fox B's ears help it survive. Explain how. [1]

4. The diagram shows a plant called a living stone which grows in the desert in South Africa. It has a long root and two swollen leaves.

   Explain how the plant's adaptations help it survive. [2]

5. Polar bears are adapted to life in the Arctic. Camels are adapted to life in the desert.

   Suggest why both animals have very wide feet. [2]

   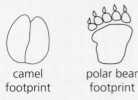

   camel footprint   polar bear footprint

6. The creosote bush grows in the North American desert. It has one long root and a web of surface roots. The surface roots secrete a chemical that stops other plants growing close to it.

   Explain how these adaptations help it survive. [2]

7. Decorator crabs cover their shells with corals, sea anemones, seaweeds, and sponges.

   Suggest how this behaviour helps them survive. [1]

8. Feathers trap air so they are good insulators, but a lot of heat escapes through a bird's beak. Scientists compared the average beak lengths of seed-eating birds from different climates.

   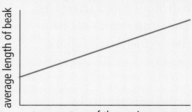

   a. What does the graph show? [1]

   b. Suggest how this correlation could help birds survive in different climates. [1]

9. Karis found two different species of plant, A and B, growing close to a tree. She used a quadrat to measure the percentage of each plant covering the ground at different distances from the tree.

   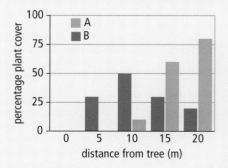

   a. Which plant appears to be better adapted to the dim light under the tree? [2]

   b. How could Karis make her results more reliable? [1]

10 The minimum area of sea ice in the Arctic has been measured for more than 30 years.

Year	Minimum area of sea ice (millions of m²)
1978	8.2
1990	6.9
2002	6.75
2008	5.4
2012	3.5

a By how much did the minimum area of sea ice change between 1978 and 2012? [2]

b Narwhals feed under sea ice. Orcas hunt in open waters all over the world. They can eat narwhals but they can't swim under ice to catch them. [2]

narwhal orca

Explain why narwhals could be endangered if the sea ice disappears.

11 Seema notices a poisonous plant growing near her school. She decides to find out how many plants there are growing in a square of grass with an area of 50 m².

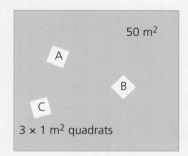

50 m²
A
B
C
3 × 1 m² quadrats

She throws a quadrat over her shoulder and counts the number of poisonous plants inside it. Then she does the same at two more positions. The table shows her results.

Quadrat	Poisonous plants
A	1
B	2
C	3

a Explain why Seema threw the quadrat over her shoulder instead of just putting it down on the grass. [1]

b Estimate the total number of poisonous plants growing in the square of grass. [1]

c How could Seema get a more reliable estimate? [1]

12 In Europe all the dangerous snakes are vipers, which have triangular heads.

Grass snakes are not poisonous. When they detect a predator such as a bird of prey, they squash their heads to make them look triangular. Scientists used model snakes to test whether this behaviour helps grass snakes survive.

They made model snakes and left them in a field for a week. Birds of prey left claw marks on them every time they attacked. This allowed the scientists to count the number of attacks on each snake.

The bar chart shows the number of attacks on each type of model snake.

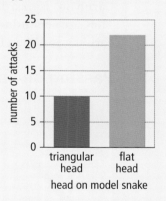

a Describe the results. [1]

b Explain how the grass snake's behaviour helps it to survive. [1]

13 Rahul lives near a lake. Last week he caught 10 fish. He kept them in a net while he marked their fins. Then he released them. This week he visited the lake again and caught another 10 fish. Only one of them was already marked.

Choose the most likely size of the fish population in the lake. [1]

10 fish   100 fish   1000 fish

15.1 Food webs

Life in the Sahara

Objective
- Construct food chains and webs and explain what they show

Camels have adapted to life in the desert. They can walk all day, surviving high temperatures and lack of water. Most smaller desert animals rest underground or in shadows during the day.

↑ Camels have adapted to life in the desert.

In the desert there are many herbivore species of insect, lizard, and rodent. They are **primary consumers**, feeding on the desert grasses, shrubs, and trees. Even when nothing shows above ground, there may be roots to feed them underground.

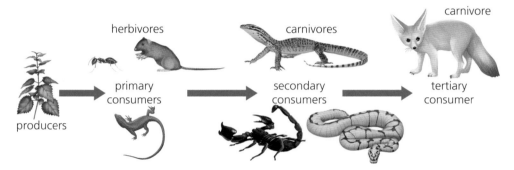

↑ All animals are consumers – they cannot make their own food.

Carnivorous lizards, spiders, snakes, and scorpions prey on the herbivores. They are **secondary consumers**. Fennec foxes eat a mixture of herbivores and carnivores. When they consume other carnivores, they are **tertiary consumers**.

Passing on energy

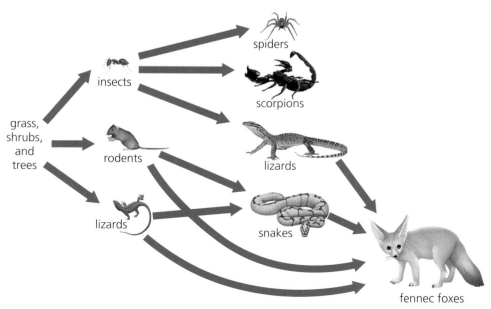

↑ Food chains link to form food webs.

Food chains link to form food webs. Most animals have more than one source of food.

In a food web each image represents not just one organism but the whole **population** of that type of plant or animal. Scientists often compare the total biomass of each population in a food web.

There may be many different species present in a food web. The arrows represent the flow of energy through the web. The producers absorb light energy and produce biomass. When consumers eat this biomass, some of the energy the producers took in passes to them.

Trophic levels

The position of a living thing in a food chain is called its **trophic level**. The producers at the beginning of a food chain represent trophic level 1. Primary consumers (herbivores) make up trophic level 2, and all the animals at higher trophic levels are carnivores.

Fennec foxes feed at the third trophic level when they eat primary consumers such as rodents. They feed at the fourth trophic level when they eat secondary consumers such as snakes. So their trophic level is somewhere between 3 and 4.

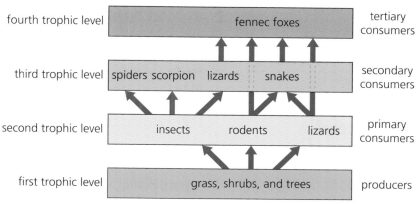

↑ Each step in a food chain represents a trophic level.

Ocean food webs

Most of the producers in the oceans are microscopic living things called phytoplankton. These grow and reproduce quickly. They support huge populations of tiny herbivores called zooplankton. These in turn feed larger invertebrates and small fish.

Small fish are eaten by squid and larger fish, and these are eaten by larger predators like sharks and dolphins.

The food webs in the ocean usually have more trophic levels than those on land.

← Ocean food webs usually have more trophic levels than those on land.

Q

1. Use the following information to construct a food web:
   Snails, woodlice, and millipedes feed on dead leaves. Beetles eat millipedes and beetle larvae eat snails and woodlice. Spiders are also carnivorous. They eat both beetles and their larvae.
2. Which of the animals in question 1 are secondary consumers?
3. Which of the animals in question 1 make up the second trophic level?
4. What types of living thing are the producers at the starts of food chains?
5. Explain what the arrows in a food web represent.

- Energy is passed along food chains from producers to consumers.
- Each step in a food chain represents a trophic level.
- Food chains link to form food webs.

15.2 Energy flow

Objective
- Model energy flow through food chains

Carnivores

Cheetahs are the world's fastest land animals, but there aren't many of them. As you go up a food chain, the animals become larger and fewer.

Cheetahs are carnivores. They feed mainly on impalas. To get enough food they need a large area of land and access to plenty of their prey. As their land is taken for towns and farms, their numbers keep dropping. Many wild carnivores are endangered.

↑ Many wild carnivores are endangered.

Pyramids of number

Food chains show what each animal eats, but they don't show how many plants or animals there are at each level.

A **pyramid of numbers** shows how many of each organism there are in a food chain. Leopards prey on impala, and impala feed on grass. A lot of grass in the first trophic level feeds a smaller number of impala in the second trophic level. These provide food for an even smaller number of leopards in the third trophic level.

Pyramids of number don't all have the same shape. When insects feed on trees, the pyramid shape is spoiled. Trees are much larger than insects, so the second trophic level contains more organisms than the first level.

↑ A pyramid of numbers is usually narrower at the top.

↑ The shape of a pyramid of numbers is distorted when the organisms in each trophic level are very different sizes.

Passing on energy

Pyramids of number show the numbers of organisms at each level. They don't show the amount of energy that is passed up from plants to predators.

When impalas eat grass, they do not get all the food the grass plants made. Plants use most of the food they make for respiration. Respiration happens all the time in every plant cell, just as it does in animals. The energy it releases keeps plants alive and lets them grow.

Impalas can only get food that is stored in a plant's leaves. Impalas can't digest some parts of leaves, and these pass out in their faeces – so they only get some of the energy that was stored in the leaves.

Just like the plants, impala use most of their food for respiration.

When a cheetah eats impala, it only gets the food the impalas stored in their muscles and fat while they were growing. This is only a tiny percentage of the energy the plants absorbed from the Sun. So cheetahs need to eat a lot of impala, and impala need to eat a lot of grass.

Energy flow

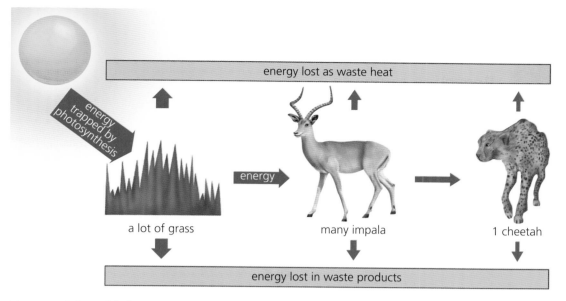

← Energy is lost at each step in a food chain.

Pyramids of biomass

Pyramids of biomass are always shaped like pyramids because they show the mass of living things in each trophic level – not their number.

Only a fraction of the biomass in each trophic level is passed to the animals that consume it. Most is used for respiration. So the total biomass in a higher trophic level is always less than in a lower one.

Energy flow

Pyramids of biomass are steeper in aquatic environments. Most of the producers there are microscopic organisms. They have fewer indigestible parts so the herbivores get a bigger percentage of the biomass from these producers, and more of the energy they absorb from the Sun is passed on. Aquatic pyramids of biomass have up to six trophic levels.

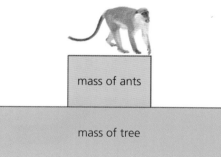

↑ A pyramid of biomass shows the mass of living things in each trophic level.

1. In the food chain below, the numbers show how many organisms there are. Draw a pyramid of numbers for the food chain. Draw the bars to scale using 1 cm width for each organism.

   Acacia trees → giraffes → lion
      15   2   1

2. Now draw a pyramid of numbers for this food chain, again using 1 cm for each organism.

   Acacia trees → impalas → cheetah
     5   10   1

3. What happens to most of the energy in the food that impalas eat?

4. Draw a pyramid of biomass for this food chain. Use 1 mm for every 100 kg.

   Acacia tree leaves → giraffes → lion
     20 000 kg   1500 kg   150 kg

5. Convert this food chain into a pyramid of biomass.

   Grass plants → impala → cheetah
     10 000 kg   500 kg   50 kg

6. There are predators and prey in most parts of the world. Which are present in the higher numbers?

7. What is the same about every pyramid of biomass?

8. Why do big carnivores such as lions need a large territory with plenty of vegetation?

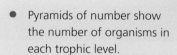

- Pyramids of number show the number of organisms in each trophic level.
- Pyramids of biomass show their mass.
- Energy is lost at each trophic level so the total biomass drops.

15.3 Decomposers

Objective
- Explain the role of decomposers

Recycling minerals

As energy flows along food chains, it gradually leaks away as waste heat. But the atoms in living things are constantly recycled.

Decomposers break down animal waste and dead plants and animals. These decomposers include fungi and bacteria (see page 31) and invertebrates such as worms, slugs, snails, and fly larvae.

As they break down the waste, decomposers return minerals to the soil. Then plants absorb the minerals and build new biomass.

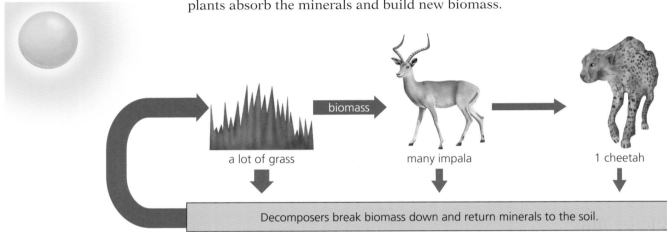

↑ Decomposers help to recycle the elements in living things.

Starting new food chains

Mangrove trees grow right at the edge of the sea along sheltered tropical coasts. A huge variety of life lives amongst the trees, and hundreds of other species use them as breeding grounds.

On land, most plants are eaten by herbivores. In mangrove forests, most leaves eventually fall into the water. Instead of being eaten, they are decomposed by bacteria and fungi.

↑ Mangroves are home to a huge diversity of species.

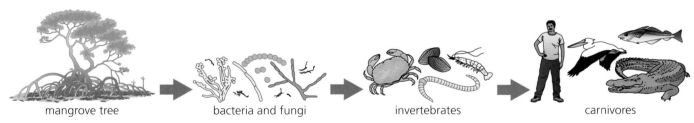

↑ The decomposers in mangrove forests form part of many different food chains.

A huge variety of invertebrates feeds on the decomposers. These invertebrates are eaten by many different carnivores. So mangrove forests have a high **biodiversity** and support complicated food webs.

Recycling carbon

The food that plants make consists of carbon compounds. Some of these carbon compounds are used for respiration. Others are passed to the next link in the food chain.

Each consumer in turn uses some carbon compounds for respiration, and others are passed along the food chain. The respiration of all the organisms in the food chain returns carbon dioxide to air and to the oceans.

Dead plants and animals are broken down by decomposers. Decomposers also respire so they also return carbon dioxide to the air.

When forests are cleared, decomposers break down the roots and branches left behind. So deforestation adds a massive amount of carbon dioxide to the atmosphere.

The importance of minerals

Mangrove forests provide building materials, fuel, medicines, and food, and the waters around them are rich fishing grounds. They produce new biomass as fast as tropical rainforests, but why are they so productive?

Plants need light, water, carbon dioxide, minerals, and a suitable temperature. If any of these is in short supply, photosynthesis slows down.

Most parts of the ocean are short of minerals. Tiny plants called phytoplankton take minerals from the water and get eaten by microscopic animals. When these die, they fall to the bottom and take the minerals with them.

Mangrove forests grow on the coast where rivers bring extra minerals down to the shore, and tides stir them up from the seabed. Because minerals are readily available, the mangroves can have high rates of photosynthesis. Mangroves take in these minerals and pass them along food chains to animals, including humans.

Unfortunately 50 % of the world's mangrove forests have been cleared in the last 50 years. Others are threatened by pollution brought down to the coast by rivers.

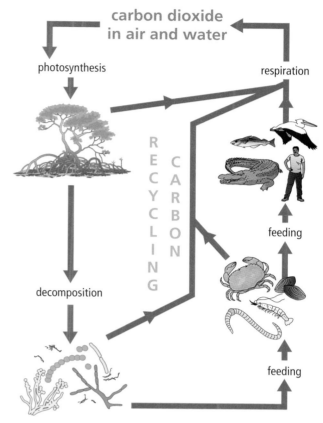

⬆ Decomposers help to return carbon dioxide to the atmosphere.

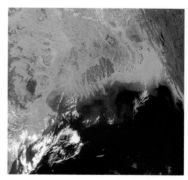

⬆ The dark area on this satellite image is mangrove forest in the Bay of Bengal.

Q

1. Name two sorts of decomposer.
2. Most of the food chains in the mangroves depend on decomposers. Explain why.
3. Why are mangrove forests important to local people?
4. Why are mangrove forests important to the Earth as a whole?
5. Decomposers break down carbon compounds in rotting leaves. Suggest two ways the carbon in these leaves could get back into the air.
6. Explain why life on Earth couldn't continue if there were no decomposers.
7. **Extension:** explain why phytoplankton grow faster in shallow coastal waters than they do in the middle of the ocean.

- Decomposers break down dead plants and animals to release energy.
- They release minerals from plant and animal wastes, and return carbon dioxide to the air.

15.4 Changing populations

Objective
- Describe factors affecting the sizes of populations

Producers and consumers

A few herbivores were taken to an uninhabited island. Their population rose.

Eventually there were so many animals that the plants could not grow fast enough to feed them all. Most of the animals died.

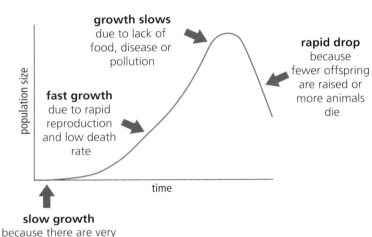

When animals move into new environments, they may find plenty of food and no predators. They can reproduce and raise offspring successfully, so their population increases rapidly.

When the population gets larger, food becomes harder to find. The population stops growing because fewer offspring are raised, or because weaker animals starve. Diseases spread more easily when animals are crowded together, so this may also cut their number.

Large populations often damage or pollute their environments. Once this happens the environment will always support fewer animals in the future.

More people, more problems

So far, nothing has stopped the human population growing. We keep changing our environment to help it support more people. Scientists disagree about whether this can continue.

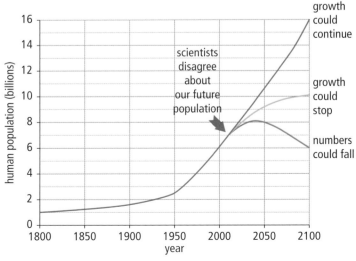

↑ In the past 200 years the human population has grown dramatically.

↑ These children work hard in school. They expect to be wealthier than their parents and have a higher standard of living.

Farming has become more efficient so we can grow more food on less land. We use machinery to process food, build homes, and improve sanitation. We have also cut the number of deaths caused by infectious diseases. But can our population carry on growing forever?

To avoid the same fate as other animals we need to make sure we don't run out of food, water, energy, or materials. We also need to prevent our environment from becoming polluted by our waste products.

Predators and prey

In habitats with low **biodiversity** – like the Arctic – a predator may rely on one prey animal. This makes the predator and prey **interdependent**. Changes in the population of one animal have a direct effect on the other animal's numbers.

In some parts of the Arctic, caribou are the main herbivores and wolves are the main carnivores. As wolves eat caribou, the numbers of caribou drop. Then wolf numbers drop too because there are fewer caribou to feed them. Weaker wolves may starve and others may raise fewer offspring.

Lower wolf numbers give the caribou a chance to breed and raise offspring. Then, as soon as their numbers increase, predator numbers increase and the cycle begins again.

↑ Caribou are prey.

Predators improve the health of prey populations because they usually catch only the oldest and weakest animals. They also stop prey numbers rising so high that they destroy their environment.

Populations are more stable in environments that have a greater biodiversity, such as forests. Bengal tigers eat several different species of deer and antelopes, bison, water buffaloes, goats, wild boar, monkeys, porcupines, hares, and birds. If one of their prey disappears they simply change their diet.

↑ The populations of caribou and wolves are interdependent.

↑ Wolves are predators.

1. Suggest why the population of an animal species might suddenly increase.
2. List three things that could reduce the population of an animal species.
3. Many animals produce a lot of offspring, but their populations stay small. What stops their numbers rising?
4. All animals need food and water. What extra resources do humans use?
5. Explain what is meant by sustainable development.
6. What does it mean if animals are interdependent?
7. Explain why the numbers of caribou and wolves in the Arctic rise and fall in cycles.
8. Explain why populations are more likely to stay the same size where there is more biodiversity.

- Population sizes are normally controlled by the amount of food available, and the numbers killed by predators and diseases.
- A species with no predators can reproduce quickly until the large numbers cause pollution, disease, or lack of food.
- Sustainable development could stop this happening to humans and preserve other species.

15.5 Facing extinction

Objective
- Describe factors affecting the size of populations

Invaders

Guam is an island in the Pacific Ocean. Sixty years ago the forests on Guam were full of birds. Then brown tree snakes reached the island, hidden in a crate of supplies. The birds began to disappear. Most of the island's birds are extinct now. The snakes ate them faster than they could reproduce.

↑ The snake population on Guam has been growing for 60 years.

The snake population on Guam has been growing for 60 years. With plenty of prey and no predators, snake numbers increased dramatically. By 2012 there were 2 million of them on the island.

In their original habitat, the snakes had to compete with other species for food. They also had predators. Most of their offspring did not survive, so their population did not grow. On Guam there was nothing to stop them.

Invasive species like the snakes on Guam usually grow quickly, reproduce quickly, survive in a wide range of habitats, eat anything, and beat other species in the competition for food and spaces to live.

Lost species

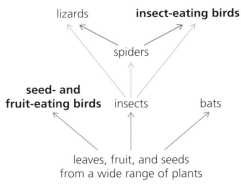

↑ Part of the food web on Guam before the snakes arrived.

↑ Guam kingfishers are extinct in the wild but a few have survived in zoos.

The birds on Guam had no predators before the snakes arrived. They had no defences and nowhere to hide.

Species are more likely to become extinct if they are only found in a few places and they reproduce very slowly. Guam's birds were wiped out by predators. Other animals become extinct when a new competitor arrives and cuts their food supply.

Scientists thought the number of snakes would drop once all the birds had gone, but their numbers kept on rising. They just swapped to eating other animals – mainly lizards.

Energy flow

Disrupted food webs

The jungle on Guam is full of spiders' webs now.

Birds no longer eat the spiders or compete with them for insects. Some spiders are eaten by lizards, but lizard numbers have fallen since the snakes arrived. So spider numbers are much higher than they were before the snakes arrived.

Scientists think its a good thing there are so many spiders. Leaf-eating insects would strip the island's vegetation if the spiders weren't there to eat them.

⬆ Now spiders have fewer predators their numbers are increasing.

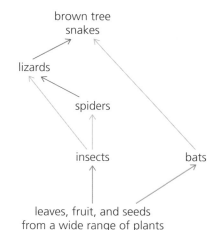

⬆ Part of the food web on Guam in 2012.

Long-term consequences

When plants and animals share habitats, they often become interdependent. More than half of Guam's tree seeds were dispersed by birds. The birds ate each tree's fruit and then dropped the seeds somewhere else in their faeces.

Now there are no birds, seeds fall under their parent trees where they can't get enough light to grow. They no longer get carried to gaps in the forest where they could replace fallen trees.

Removing the snakes

In the past, people have removed invasive species from other islands. The methods used include: killing them individually; destroying their habitats; using a predator or a disease to kill them; interfering with their reproduction; or destroying them with toxic chemicals.

⬆ Birds no longer carry tree seeds to fill gaps in the forest like this.

The snakes on Guam are difficult to kill because they hide in the treetops. Pest control experts keep trying different ways of killing them. In one attempt they parachuted dead mice full of poison into the treetops. The parachutes stopped the mice falling to the forest floor where other species would have grabbed them.

Q

1 List the characteristics that make invasive species successful.
2 What characteristics make a species more likely to become extinct?
3 When the bird population on Guam dropped, the number of tree snakes still continued to rise. Why was this?
4 Use the island's food web to explain why the number of spiders on Guam has risen.
5 Some people want to destroy the spiders so tourists are not frightened away. Use the food web to explain why destroying the spiders could ruin the island.
6 Explain why the forests on Guam are likely to change in the future.

- Population sizes are normally controlled by the amount of food available, and the numbers killed by predators and diseases.
- Changes in a habitat such as new predators, or changes in the food supply, can make species extinct.

Extension 15.6

Maintaining biodiversity

Objective
- Explain why biodiversity is important

Comparing biodiversity

Tourists visit Africa to see wild animals that are found nowhere else on Earth.

Some countries have many wild plants and animals. Others have just a few. Where there are many different species, we say there is a high **biodiversity**.

Each part of the world has a different climate and receives different amounts of sunlight and rain. Each **community** of living things interacts with its environment to form an **ecosystem**.

↑ Tourists pay to observe mountain gorillas in Africa.

Biodiversity is highest in warm, wet ecosystems close to the equator. They receive more intense sunlight and get more rain, so photosynthesis takes place at the fastest possible rate. The plants in these ecosystems produce more biomass per square metre than plants elsewhere, so they can feed more animals. The plantsy rely on these animals for pollination, seed dispersal, and the recycling of minerals.

The importance of forests

The biodiversity of an ecosystem also depends on the type of vegetation it contains.

Scientists compared a tropical rainforest with other hot, wet places. They found most biodiversity in the forest.

Hot, wet ecosystems	Biomass produced (g/m²/year)	Biodiversity
rainforest	2500	very high
grassland	1000	high
farmland	600	low

Each rainforest plant needs slightly different growing conditions. The tall trees grow fast to reach above the others. They need bright light. The plants on the forest floor have darker leaves. They absorb the dim light that filters through the trees above them. So, between them, the forest plants use more of the sunlight. It's the same for every other resource such as water.

The huge range of plants in a tropical rainforest creates many different habitats for animals, so so the animals that inhabit rainforests belong to many different species.

When forests are cleared for farming, biodiversity plummets. This biodiversity is easy to lose and impossible to replace.

↑ Biodiversity drops when trees are removed to grow crops or keep animals.

Energy flow

Saving seeds

Scientists all over the world send seeds to the Millennium Seed Bank in England. Scientists here aim to preserve 25% of the world's plant species by storing their seeds.

Most of the food we eat comes from about 20 different crops. If diseases attack these crops, millions of people could starve. When a virus began to destroy rice crops in the 1970s, scientists looked for a variety of rice that could resist the disease. They tested 6273 varieties and found one resistant one. It was used it to create a hybrid plant that the virus could not destroy.

When a useful variety exists, scientists can use it to improve crops. If lots of plants disappear, that won't be possible.

As well as food, plants provide other useful products. More than half our medicines were originally found in plants and scientists predict that we could discover many more if the plants survive.

↑ Seeds from 10% of the world's species are stored here.

Conserving animals

Scientists use many different strategies to prevent animals becoming extinct (see page 60). The best method is to preserve their habitat but that isn't always possible.

When animal numbers fall too low, they can only breed with close relatives. This is called **inbreeding**. It reduces variation and cuts the population's resistance to diseases. Scientists can help animals avoid extinction by storing sperm from wild animals at very low temperatures. The sperm can be sent to other parts of the world to allow animals in zoos and sancturies to reproduce.

↑ Sperm from a wild cheetah allowed this zoo animal to reproduce.

Q
1. What does it mean if an ecosystem has a high biodiversity?
2. Explain why we should limit the amount of forest cleared for farming.
3. The trees in a rainforest can absorb more of the sunlight that reaches it than crops on farmland can. Why is this?
4. Give one reason why it is useful to have many different varieties of each crop plant.
5. In the past, animals in small populations could only breed with each other. Explain how they can now breed with animals from other countries.

- Ecosystems that contain many different species have a high biodiversity.
- Biodiversity is highest in close to the equator in tropical rainforests.
- We can conserve plant biodiversity by storing seeds.
- Animal biodiversity can be conserved by storing sperm from an animal species and using it to help isolated populations to breed successfully.

Review 15.7

1. In the desert, insects eat plants and foxes eat lizards. Most of the insects are eaten by lizards.

   a Draw a sketch that links these animals to form a food chain. [1]

   b How would the number of lizards change if the foxes all died? [1]

2. This food chain is found in the Antarctic:

   Phytoplankton → krill → squid → penguin → seal

   Name an organism from the food chain that fits each of these categories:

   a predator [1]
   b prey [1]
   c herbivore [1]
   d tertiary consumer [1]
   e primary consumer [1]
   f in the first trophic level [1]
   g in the highest trophic level. [1]

3. The diagram shows a food web in the sea around an island.

   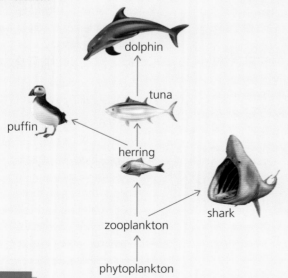

   a Fishing boats have decreased the number of herring in the sea. Explain why that could affect the number of offspring puffins can raise. [2]

   b Which animals have a bigger food supply now there are fewer herrings? [1]

   c Global warming is expected to reduce the biomass of phytoplankton in the oceans. What effect will that have on the rest of the food web? [1]

4. The diagram shows the food web in one part of a forest.

   a Which organism is both predator and prey? [1]

   b Which organism's population will have the smallest biomass? [1]

   c Explain why the biomass of secondary consumers is always lower than the biomass of primary consumers. [2]

5. Which of these statements about food webs are correct? [3]

   a Producers form the first trophic level.
   b Some food chains start with a primary consumer.
   c Most food chains rely on the Sun as the source of their energy.
   d Herbivores form the second trophic level.
   e Energy is transferred from tertiary consumer to producer.

6. The diagram below shows a pyramid of biomass.

   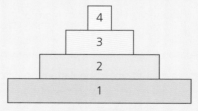

   Match each of the following organisms to the correct trophic level:

   a insect [1]
   b frog [1]
   c snake [1]
   d grass [1]

7 The diagram shows pyramids of biomass for two human food chains.

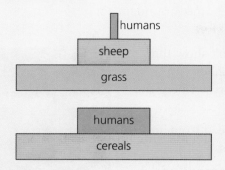

Explain why the same area of land can feed more people if they eat cereals instead of meat. [3]

8 Six rabbits were placed on a small island. The graph below shows how their population changed.

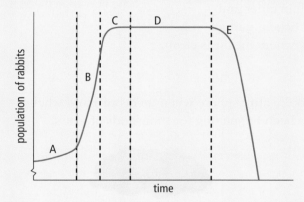

a Give the letter of the part of the graph where the number of rabbits is increasing fastest. [1]

b Which section of graph shows when the rabbits are dying as fast as they are being born? [1]

c What does part E of the graph show? [1]

d Suggest two things that could cause the change seen in section E. [1]

9 The diagram below shows how minerals are recycled.

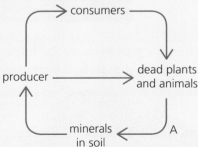

a Name the process represented by arrow A. [1]

b What type of organism carries out process A? [1]

c Name two examples of this type of organism. [2]

d Explain why process A is vital to life on Earth. [2]

10 The diagram below shows how carbon is recycled by living things.

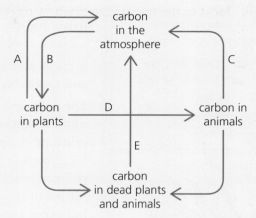

Choose the correct arrow to represent each of the following:

a photosynthesis [1]

b animal respiration [1]

c feeding [1]

d the action of decomposers. [1]

11 Phytoplankton are microscopic organisms They are found in every ocean. Like plants, they make their own food by photosynthesis. Organisms called zooplankton feed on them.

The graph shows how the biomass of each organism rises and falls.

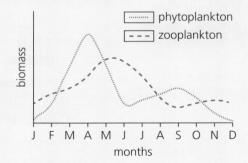

a Suggest why the zooplankton biomass decreases between June and July. [1]

b Suggest two factors that would limit the biomass of phytoplankton if there were no zooplankton eating them. [2]

16.1 Air pollution

Objective
- Describe the causes and consequences of air pollution
- Compare evidence from first-hand experience with secondary sources
- Look critically at sources of secondary data

Indicators

The leafy **lichens** on this tree trunk show that the air is here slightly polluted. Lichens are **living indicators**. More species grow where the air is cleaner.

Lichens absorb everything they need from air and rainwater. They have no roots.

The pollutants that affect lichens are **sulfur dioxide** and **oxides of nitrogen**. They make rainwater very acidic, which makes it difficult for lichens to grow. Where the rain is most polluted, only a few thin, crusty lichens grow. Air pollution can get very high in cities and industrial areas. Taller, bushy lichens cannot survive there – they grow only where the air is clean.

⬆ The variety of lichen species growing on walls and trees shows how clean the air is.

bushy lichen in clean air

leafy lichen in slightly polluted air

crusty lichen in heavily polluted air

Acidic gases

Sulfur dioxide is released into the atmosphere when fuels burn and when metals are extracted. This has been happening for thousands of years.

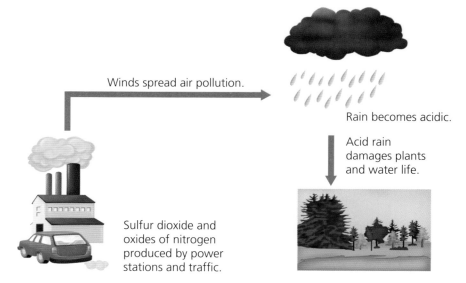

Winds spread air pollution.
Rain becomes acidic.
Acid rain damages plants and water life.
Sulfur dioxide and oxides of nitrogen produced by power stations and traffic.

We produce more pollution now than in the past because the human population is larger. More people use more materials and burn more fuels.

Traffic is the main source of acidic gases in cities. Cars release oxides of nitrogen.

Acidic gases cause breathing difficulties, which can be fatal in people with damaged lungs or asthma. Many cities release daily pollution reports. People with damaged lungs should avoid badly polluted areas.

Sulfur dioxide and oxides of nitrogen are both very soluble. They dissolve in rainwater to make **acid rain**.

Human influences

Acid rain causes a variety of problems:

- It makes soil more acidic. This means that plants can't take in essential mineral elements from the soil.
- It damages decomposers, so they don't return minerals to the soil.
- It dissolves aluminium from rocks, which can poison plants and water life.
- It damages leaves and seeds, so pathogens can infect plants more easily.
- It slowly dissolves any structures made of chalk, limestone, or marble.

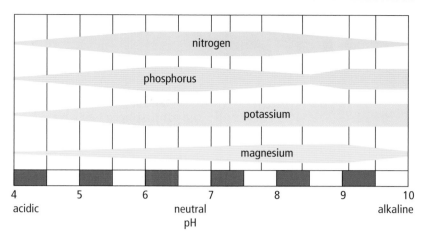

⬆ The narrower the bars, the less of the mineral element is available to plants from the soil.

Reducing pollution

Scientists try to stop pollution and repair the damage humans have done to the environment.

Most of the sulfur dioxide released by power stations can be captured before it escapes. The process is called **flue gas desulfurisation**.

Most of the oxides of nitrogen from petrol engines can be removed using catalytic converters.

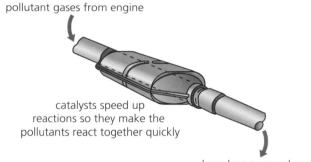

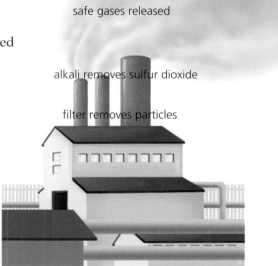

⬆ We can stop sulfur dioxide escaping when coal is burned in power stations.

Scientists collect data to see how much pollution the air contains. Each group usually takes measurements in one part of the world on specific dates. To spot global trends we need to use results collected by many different scientists. These are secondary sources of evidence.

Scientists publish their results in scientific journals. They must explain how they collected their data so that other scientists can see how trustworthy their results are.

Evidence from most countries shows that the amount of sulfur dioxide released is falling.

- Sulfur dioxide is released when fuels are burned.
- Oxides of nitrogen are released from car engines.
- Sulfur dioxide and oxides of nitrogen dissolve in rainwater and make it acidic.
- Lichens are living indicators of air pollution.
- Only a few lichen species can grow where sulfur dioxide concentrations are high.

Q

1. Name the gases that cause acid rain and explain where they come from.
2. Why can lichens be used as living indicators of air pollution?
3. How can acid rain damage plants?
4. Explain why air pollution has increased in the past 100 years.

Enquiry 16.2 How scientists work

Creative thinking

Burning fossil fuels releases carbon dioxide. The amount of carbon dioxide in the atmosphere is rising. Should we be worried about this?

Scientists started to answer this question 200 years ago. Joseph Fourier asked: *'What keeps Earth warm?'* Then he used creative thought to find an answer, *'It must be the atmosphere.'* The way the atmosphere traps heat is called the **greenhouse effect**.

John Tyndall invented apparatus and did experiments. He measured the amount of heat different gases could absorb. *Small amounts of carbon dioxide trapped a lot of heat*.

Could the greenhouse effect get bigger? Svante Arrhenius analysed the evidence and made a prediction: *Doubling the amount of carbon dioxide in the atmosphere would make it 5 °C warmer*. We could easily double the amunt of carbon dioxide by burning fossil fuels.

Objectives

- Understand the relationship between carbon dioxide production and global warming
- Recognise how scientists work today and how they worked in the past

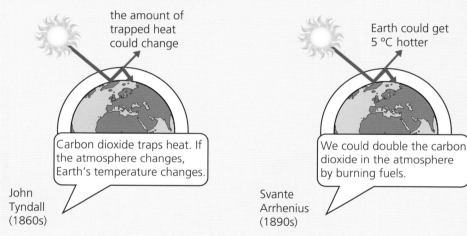

Joseph Fourier (1820s): Earth is surprisingly warm. The atmosphere must stop heat escaping.

John Tyndall (1860s): Carbon dioxide traps heat. If the atmosphere changes, Earth's temperature changes.

Svante Arrhenius (1890s): We could double the carbon dioxide in the atmosphere by burning fuels.

Investigation

↑ The observatory, Hawaii.

Observatory director says:

'Carbon dioxide is a very small part of the atosphere, so we need to track it very accurately. Even small changes in carbon dioxide have a big effect on the atmosphere.'

By the 1950s, many scientists worked for government organisations. They wanted to understand the atmosphere. Experts worked together to find out more. They used new measuring instruments to collect evidence all over the world.

One group checked the prediction Arrhenius had made. They already knew the atmosphere's temperature was rising. But was this **global warming** caused by carbon dioxide?

They set up an observatory on a mountain to measure the carbon dioxide concentration in the air.

Human influences

Correlation

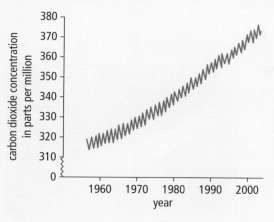

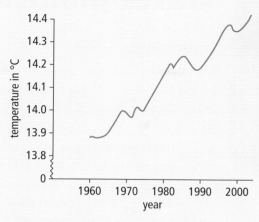

⬆ Tree rings are thicker in cool wet years.

Scientists found the correlation they were looking for. The amount of carbon dioxide in the atmosphere is increasing and so is its temperature. But that still does not prove that one factor causes the other.

In the 1980s, scientists looked back in time. They analysed **tree rings** and **ice cores** to see how the temperature, and the carbon dioxide in the atmosphere, varied in the past. They found the same correlation. The values of both variables have risen and fallen in the past, but now they are both rising faster than ever before.

each layer in an ice core contains tiny bubbles of trapped air.

Conclusions

Scientists check each other's results and argue about the conclusions. Some don't agree that carbon dioxide causes global warming. They say that natural events are more important – like changes in Earth's orbit.

⬆ The deepest layers of ice are the oldest. Scientists can meaure the carbon dioxide in the trapped air.

So is carbon dioxide to blame? Governments need answers. They set up the Intergovernmental Panel on Climate Change (IPCC), led by Indian scientist Rajendra Pachauri.

The IPCC studies evidence from thousands of scientists from all over the world. In 2007 they concluded that global warming is definitely happening. It will cause serious disasters – more storms, droughts, and heatwaves. Carbon dioxide from human activity is 'very likely' the cause.

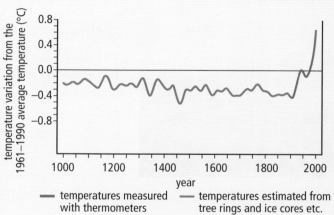

The report convinced some governments to take action to prevent **climate change**, but will their actions be effective? Only time will tell…

Q
1. How could carbon dioxide make Earth warmer?
2. What prediction did Arrhenius make about carbon dioxide?
3. Describe one piece of evidence that supports this prediction.
4. Do all scientists agree that humans cause global warming?
5. How might the weather change as the Earth warms?
6. How does the IPCC help governments to decide what to do about global warming?

- Carbon dioxide is released when fuels are burned.
- Scientists used creative thought, experimentation, and analysis of evidence to link carbon dioxide levels global warming.

16.3 Water pollution

Indicators

A few weeks ago fish were swimming in this lake. Now they cannot survive here. The lake is too polluted. How did it get like that?

The fish are dying because the water lacks oxygen. Under the thick green layer of algae the lake is full of micro-organisms. Their respiration uses up oxygen faster than plants can replace it.

↑ This fish died from a lack of oxygen.

The bacteria in the lake are decomposers. Their population is growing rapidly because they have piles of dead plants to break down. It all started when rain washed fertiliser into the water.

Algae usually grow quite slowly in lakes and rivers. The fertiliser changed that. The water's mineral content shot up and algae multiplied very rapidly. Consumers couldn't eat them fast enough. They formed thick green mats on top of the water.

The plants below the surface were suddenly cut off from light. They could not photosynthesise, so they died.

The sequence of events in the flow chart is called **eutrophication**. The lake will take a long time to recover.

Sewage

Any substance that adds minerals to water can cause pollution. The waste from our toilets acts like a fertiliser, so large amounts of untreated sewage should never be allowed to enter rivers.

Sewerage systems store waste water in huge tanks while micro-organisms break its contents down. That makes the water clean enough to send to rivers or the sea.

Sewerage systems are expensive. Many countries can't afford to treat all the waste from their cities. Large volumes of untreated sewage pollute their rivers.

Monitoring rivers

Scientists can measure the amount of pollution in a river, but it is easier to use living indicators. These can be algae, fish, or invertebrates. Invertebrates are usually used because they are easy to catch and easy to identify.

The first step is to find a relationship between the concentration of pollutants, the

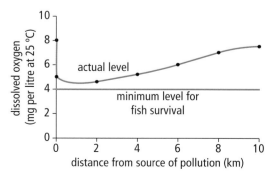

↑ Rivers can recover if their oxygen levels don't drop too low.

Objective
- Describe the causes and consequences of water pollution

fertiliser enters water

algae multiply rapidly

plants below the surface die

micro-organisms use up oxygen as they decompose dead plants

fish die

↑ This flow chart shows what can happen when fertiliser runs into a lake.

↑ Sewerage works use micro-organisms to break down human wastes.

Human influences

oxygen concentration in the water, and the species present. Then images of the invertebrates can be used to construct a pollution chart that anyone can use.

Pollution	High	Moderate	Low	None
Main species present	rat-tailed maggot and sludge worms	water louse	freshwater shrimp	mayfly nymph
Oxygen (mg per litre at 25 °C)	0–3	3–5	5–6	6–8

↑ This pollution chart was used to monitor pollution levels in UK streams.

Persistent organic pollutants

Persistent pollutants don't break down naturally, so they spread around the world in air and water.

The first insecticides had persistent molecules. They are not used any more, but they are still causing problems. They got washed into rivers, ran into the sea, and were carried around the world by ocean currents.

Fish absorb the insecticides from sea water and store them in their fat. When seals eat fish the insecticide goes into their bodies. It gradually builds up because each predator eats a lot of prey. The pollutant in each prey animal builds up in the seal's body. This is **bioaccumulation**.

Polar bears eat a lot of seals, so their bodies contain even more insecticide. Eventually, the amount in their bodies can get high enough to poison them.

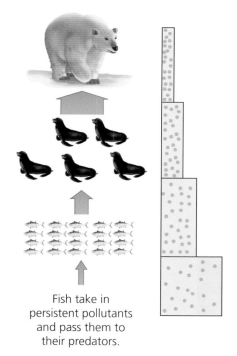

Fish take in persistent pollutants and pass them to their predators.

← Persistent pollutants become more concentrated as they move along food chains.

Q

1. Explain why fertilisers make plants and algae grow faster.
2. List the sequence of events that leads to eutrophication.
3. How can we stop sewage polluting rivers?
4. Explain how invertebrates can be used to monitor water pollution.

- Water can be polluted by fertilisers or sewage.
- Decomposers reproduce quickly in polluted water and lower its oxygen content.
- Low oxygen levels in water kill fish and many invertebrates.
- Indicator species can be used to monitor pollution.

16.4 Saving rainforests

Objectives
- Understand why deforestation is happening
- Discuss ways of limiting deforestation

Landslide

Heavy rains washed a hillside onto this road. Elsewhere whole villages have been buried in mud.

Disasters like this are common where lots of trees are cut down or burned. This **deforestation** weakens hillsides.

↑ Heavy rains can wash bare soil away.

Tree roots act like giant nets. They hold the soil in place. At the same time, their leaves shelter the soil from strong sunlight and heavy rain. When the trees have gone, heat makes bare soil dry and loose. Then wind and rains can carry it away.

Deforestation

People remove trees for many reasons:
- to make space to grow food or keep farm animals
- to remove metal ores, oil, or coal from the ground
- for wood to use as a fuel or building material
- to grow food crops to sell to other countries
- to grow biofuels to sell to other countries
- to build new roads and houses.

↑ The trees here were removed to make room to grow crops.

Clearing forests damages the environment. Plants and animals lose their habitats and people lose the food and fuel the forest provided. Without trees, the local climate can become warmer and windier, which dries the soil. Bare soil is easily washed away. The lost soil fills river beds and makes floods more common.

Deforestation affects every country. As the number of trees falls there is less photosynthesis. Less carbon dioxide is removed from the atmosphere and global warming increases.

Making forests more valuable

In forests that are managed to be sustainable, trees are planted to replace the ones that are cut down. But less than 10% of rainforests are managed like this. Rising food and fuel prices make it more likely that forests will be cleared to grow crops.

Saving tropical forests is an international responsibility. Countries with no forests of their own can help to save tropical forests. They need to make them more valuable to the people who live in them and the countries that own them. These are some of the things that governments and charities are trying to do:

- encourage tourists to visit the jungle, to provide local employment
- harvest rainforest products such as fruits and nuts for export
- prevent poaching and illegal logging by outsiders

↑ Tourists stay at this forest hotel and pay local people to guide them through the forest.

- pay people to plant trees on land that is unsuitable for crops
- encourage wealthy people to buy trees and protect them.

Sustainable forests can provide a lot of work. Local people are employed to cut down trees for sale and plant new ones, and tourists spend money in the area. If countries can earn enough from the forests, they can afford to save them to benefit the rest of the world.

Reducing demand for energy and resources

Wealthy countries can also help by using fewer resources. To cut the resources they use, they are encouraged to:

- limit population growth
- reuse products and recycle materials
- use energy more efficiently so less fuel is needed
- make farms more efficient so less land is needed for crops.

Reducing our use of resources also cuts the amount of energy needed to make things and reduces pollution.

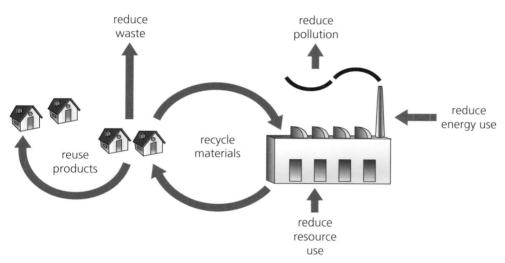

↑ Reuse and recycling cut demand for resources and energy and reduce pollution.

Q

1. Give three reasons why a growing population has to remove trees.
2. Explain how other countries encourage tropical countries to remove forests.
3. Describe three ways in which deforestation damages the local environment.
4. How does deforestation affect the world's climate?
5. Describe two ways in which countries can earn money from sustainable forests.
6. Explain how wealthy countries can reduce the need to clear forests.

- The removal of trees from forests is deforestation.
- It gives growing populations an income and extra space.
- The loss of trees destroys habitats and damages the environment.
- Forests can attract tourists and bring economic benefits.
- We can reduce the need to clear forests by using fewer resources and less energy.

Review 16.5

1 The diagram shows four common sources of pollution.

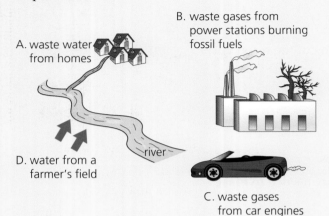

- A. waste water from homes
- B. waste gases from power stations burning fossil fuels
- C. waste gases from car engines
- D. water from a farmer's field

a Which two can cause air pollution? [2]

b Which two can pollute water? [2]

c Name the acidic gases in sources B and C. [2]

d Name the main pollutants in sources A and D. [2]

e Which two sources release most carbon dioxide? [2]

f Suggest why the trees behind the factory are dying. [1]

2 In 1952, UK scientists found a correlation between sulfur dioxide levels in the air and the number of deaths per day.

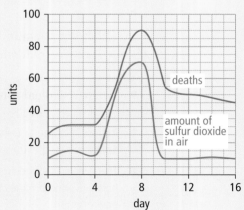

a How do humans add sulfur dioxide to the air? [1]

b Many doctors think that sulfur dioxide harms human health. Explain how these graphs support this idea. [1]

c Which part of the body does sulfur dioxide damage? [1]

d Why is reducing acid rain an international problem? [1]

3 The graph shows the number of lichens growing at different distances from a city centre.

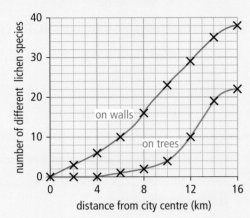

a Which surfaces have more lichen species growing on them, walls or trees? [1]

b How does the distance from the city centre affect the number of different lichen species found? [1]

c Lichens are an 'indicator species' because they cannot grow in very polluted air. Name the main source of air pollution in cities. [1]

d Where are tall, bushy lichens most likely to be found? [1]

4 The table below shows some invertebrates and the amount of oxygen they need to survive.

Invertebrate	Oxygen needed
rat-tailed maggot	very little
leech	a little
caddis fly larva	a moderate amount
stonefly nymph	a lot

Students collected water from three different parts of a river and identified the most common invertebrate present.

Water sample	Invertebrate present
A	leech
B	stonefly nymph
C	rat-tailed maggot

Use the data to list the water samples in order from least to most polluted. [3]

5 A student tested water samples from two different parts of a river. The bar charts show his results.

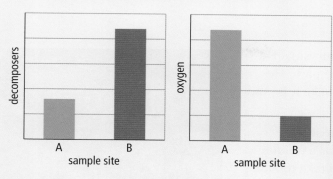

   a Which part of the river is more polluted? [1]

   b Explain why the oxygen concentration is different at the sample sites. [1]

6 Scientists investigated a river. The graph compares the mass of fish and of bacteria found at different distances from the centre of a city.

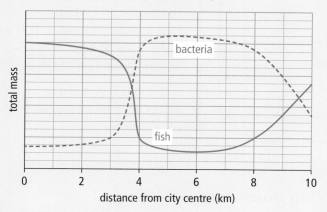

   a Suggest why the mass of bacteria increases 2 km from the town centre. [1]

   b Explain how the extra bacteria could reduce the number of fish in the river. [1]

7 The table shows the minimum oxygen levels that different fish need in the water.

Species	Oxygen needed (mg per litre)
tench	0.7
roach	0.8
perch	1.2
trout	3.7

Which species would survive best in water polluted with sewage? [1]

8 Some villagers plan to remove the trees from this hillside.

Doing this could damage their environment.

   a Suggest one immediate problem for the animals that live on the hill. [1]

   b Suggest one problem that could arise during the rainy season. [1]

9 Scientists estimate how much carbon dioxide different activities produce.

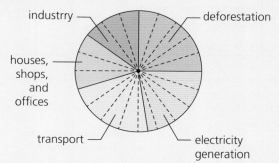

   a What percentage of carbon dioxide emissions come from industry? [1]

   b Describe two ways in which deforestation increases the amount of carbon dioxide in the atmosphere. [2]

10 In South America, huge areas of tropical forests are being cleared. The wood is sold and crops such as coffee are planted instead.

   a Give two reasons why this damages the global environment. [2]

   b Suggest two ways of limiting the damage to the forests. [2]

17.1 Using keys

Changing ideas

Victor is out in the countryside. He sees a snake lying on the path. He wonders what type of snake it is, and whether it is dangerous.

There are 2700 different snake species. Each species has a unique set of features. It would take too long to look through pictures of every snake. Instead, Victor uses a key. It quickly tells him what species the snake belongs to.

Objective
- Use and construct keys

↑ The snake has hollow fangs and a heart-shaped head. Could it be dangerous?

The snake's hollow fangs, its heart-shaped head, and the pattern on its skin show that it is *Bitis arietans*. It is a dangerous African adder, so Victor keeps away from it.

A key can be set out as a series of branches, or as a set of questions.

1	Hollow fangs drop down when it opens its mouth	see 2
	No hollow fangs	see 6
2	Heart-shaped head	see 3
	Head not heart shaped	see 5
3	Snake has 18–22 V-shaped stripes	*Bitis arietans*
	Snake has a different pattern	see 4

↑ Part of a numbered key.

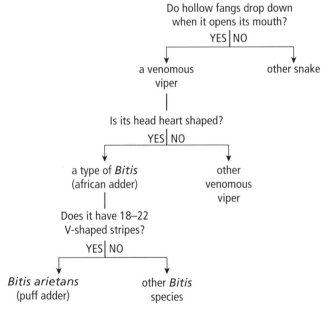

↑ Part of a branching key.

Asking questions

Keys use questions to eliminate choices until only one species is left. The questions need to be about features that don't change as an organism gets older. Features such as length, mass, and exact colour are not so helpful because they change as an animal grows.

Keys are easier to construct if both sexes have the same features, but some male and female animals look very different.

Designing keys

Victor has photographs of four African cat species. He constructs a key to help his friends identify them.

Victor's first question has to be very general. It needs to use an obvious feature to split all the cat species into two main groups. He decides to ask, "Does it have spots?".

Variation and classification

Further down his key he needs to use more subtle differences to separate the animals in each group. All his questions need to be based on facts, not opinions. A question such as, "Does it look strong?" is not useful because people may not all agree on the answer.

Victor uses the question, "Are its spots in groups or separate?" to distinguish between the leopard and the cheetah. Then he asks, "Are its ears rounded or pointed?" to separate the lion and the caracal.

↑ Four African cat species.

leopard spots arranged in rosettes

cheetah separate spots

cats with spots

cats with no spots

Using extra evidence

Closely related species can look very different, and unrelated species can look very similar. DNA evidence shows that lions and leopards are more closely related than lions and caracals or leopards and cheetahs. One feature that lions and leopards share is that they are bigger than other cat species, and they roar. Other cat species are smaller and they purr.

lion rounded ears

caracal pointed ears

large cats that roar

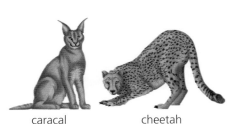

smaller cats that purr

Q

1. Use Victor's questions about cat species to make a numbered key to identify lions, cheetahs, leopards, and caracals.
2. Turn the numbered key you made in question 1 into a branching key.
3. Tigers are large cats with stripes. Modify the banching key you made in question 2 to include them.
4. **Extension:** tigers roar. Make a new branching key for the same animals that begins with the question: "Does it roar?".

- Keys use a series of questions to distinguish each species from all the rest.

17.2 What makes us different?

Identical twins

Priyanka and Juvina are identical twins. They were separated at birth and raised by different families. At birth, they were impossible to tell apart. Their faces are still very similar, but Priyanka is taller and slimmer than her sister. Why were they identical to begin with, and what has changed in the years they spent apart?

Objective
- Understand that organisms inherit characteristics from their parents through genetic material that is carried in cell nuclei

↑ These young babies are identical twins.

Inherited features

Priyanka and Juvina inherited some of their characteristics from each of their natural parents. Some of their characteristics were present when they were born, such as their skin colour and blood group. Others developed later, like the sounds of their voices and their final height.

The twins' lives began when one of their father's sperm cells fertilised one of their mother's eggs. The nucleus of the sperm cell joined with the egg cell nucleus. Then the fertilised egg began to divide. It made more cells and formed an embryo.

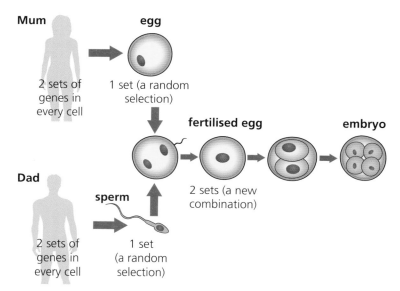

↑ A fertilised egg contains genes from each parent.

Genes

The nucleus of every cell contains a set of instructions called **genes**. Genes control cells so they control growth and development. Some genes are the same in everyone, so we all have similar bodies. Other genes help to make us all slightly different

Most cells have two copies of every gene, but egg and sperm cells have only one copy of each gene. The genes from the father and mother combine during fertilisation, so each fertilised egg has a new combination of genes.

A fertilised egg cell divides into two, then four, and so on. All the new cells in an embryo are made this way.

Before a cell divides, it copies its nucleus and all the genes inside it. So all the cells in an embryo have an identical set of genes. These genes control each cell, and so they control the characteristics of the embryo as a whole. Each fertilised egg contains a different combination of genes, so each embryo is different.

Identical twins

Priyanka and Juvina are identical twins because they grew from the same fertilised egg. When their embryo was just a bundle of cells, something made it split in two. The two groups of cells had identical collections of genes so they grew into identical baby girls.

Other influences

Genes do not change, but the twins are not completely identical now they are older.

The twins both have the same blood group. This characteristic is controlled by genes and nothing else. Other characteristics are affected by a wide range of environmental factors such as our diet, health, activities, and surroundings. These environmental factors modify the characteristics that genes produce.

Priyanka and Juvina grew up in different families so different environmental factors modified their characteristics in different ways. Priyanka always ate a healthy diet and took regular exercise. Juvina ate a lot of fatty and sugary foods when she was growing up. She didn't exercise and was ill for a long period of time. Now she is not as tall as Priyanka, but she is heavier.

↑ These women looked identical when they were younger.

Genes and environmental factors work together to produce characteristics such as height, body mass, and intelligence.

Language

Juvina and Priyanka find it hard to communicate. Priyanka's adopted family brought her up speaking Tamil. Juvina's family spoke Hindi. Our genes make it possible for us to learn to speak, but the language we use is determined by the environment we grow up in.

Q

1. Explain how an embryo gets a full set of genes.
2. How do genes you inherit from your parents get into every cell in your body?
3. Many parents have more than one child. Explain why all their children are different.
4. Why do identical twins look the same when they are young?
5. **Extension:** name one human characteristic controlled only by genes.
6. **Extension:** give two examples of characteristics influenced by both genes and environmental factors.
7. **Extension:** name one human characteristic controlled only by their environment.

!

- We inherit some characteristics from each parent because we inherit genes from them.
- Genes influence most of our characteristics by controlling our cells.
- A copy of the genes we inherited is stored in the nucleus of every cell.
- Each egg and sperm cell contain half of each parent's genes.
- Each fertilised egg contains a unique combination of genes.

Extension 17.3 Chromosomes

Objectives
- Recognise that genes are parts of chromosomes
- Understand that we inherit two copies of each chromosome, one from each parent

Inside cells

Cell nuclei usually look like a dark circles under the microscope, but special stains make them show up more clearly.

When a cell is ready to divide, long thin threads appear in its nucleus. They split into two identical sets and move to opposite ends of the cell. Each set forms a new cell nucleus as the original cell divides.

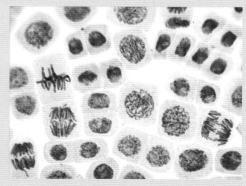

↑ The cells in this light microscope image have been stained to make their nuclei show up more clearly.

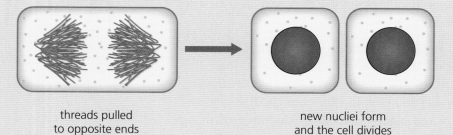

threads pulled to opposite ends of the cell

new nucliei form and the cell divides

↑ A dividing cell copies everything in its nucleus before it splits into two new cells.

Genes and chromosomes

The long thin threads in cell nuclei are called **chromosomes**. They are made from a giant molecule called **DNA**. Scattered along each chromosome are special sections of DNA called **genes**.

Once a cell has finished dividing the chromosomes are hidden inside the nucleus. The nucleus in the diagram has two pairs of chromosomes, but real human cells have 23 pairs. We inherit 23 chromosomes from each parent – one chromosome in each pair comes from each parent. So we inherit half of each parent's genes.

The chromosomes in the diagram each contain four different genes, but real human chromosomes contain up to 1000 genes.

The genes on paired chromosomes do the same jobs as each other, so we inherit two copies of each gene – one from each parent.

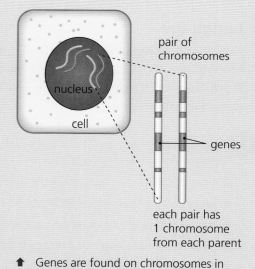

↑ Genes are found on chromosomes in the nucleus of every cell.

Sex chromosomes

One pair of human chromosomes comes in two forms called X and Y. These chromosomes carry the genes that decide whether you are male or female. Females have two X chromosomes and males have an X chromosome and a Y chromosome.

Boy or girl?

Each human egg or sperm cell contains 23 chromosomes – one from each pair.

Egg cells always contain X chromosomes but males can make two sorts of sex cells. On average, half a man's sperm carry X chromosomes and half carry Y chromosomes.

Sperm containing X or Y chromosomes have an equal chance of fertilising egg cells. So a fertilised egg cell could get two X chromosomes and produce a girl, or get an X and a Y chromosome and produce a boy. The diagram shows that the chances of producing a boy are 2 in 4, which is the same as 1 in 2 or 50%.

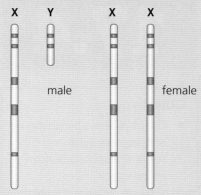

↑ One pair of chromosomes is different in males and females.

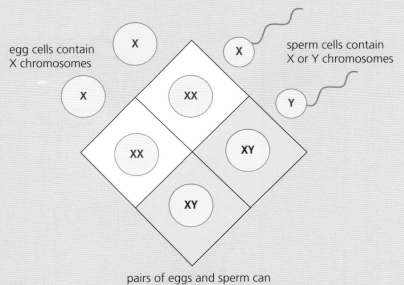

↑ On average, equal numbers of fertilised eggs produce males (XY) and females (XX).

- Genes are parts of chromosomes and are made of DNA.
- The nucleus of a human cell contains 23 pairs of chromosomes.
- One of the chromosomes in each pair was inherited from each parent.
- Females have two identical sex chromosomes (XX).
- Males have two different sex chromosomes (XY).
- Eggs and sperm each contain one chromosome from each pair.
- Fertilised eggs are equally likely to produce males or females.

Q
1. All your cells contain a copy of the genes you inherited. Draw a diagram to show where these genes are kept.
2. How many chromosomes do your body cells contain?
3. Explain why you have two copies of every gene.
4. How many chromosomes do your sex cells contain?
5. What decides whether a fertilised egg grows into a boy or a girl?
6. Explain why parents are equally likely to have a girl or a boy.

Extension 17.4

Investigating inheritance

Objective
- Use ideas about dominant and recessive genes to predict the characteristics offspring will inherit

Gregor Mendel

A monk called Gregor Mendel used creative thinking to explain inheritance 150 years ago – before genes were discovered.

Mendel worked with tall and short pea plants. Each pea flower makes male and female sex cells, so they can self-pollinate and produce tall or short offspring identical to the parent plant.

↑ Mendel's work helped to explain how characteristics are inherited.

Mendel stopped flowers self-pollinating by taking away their anthers. Then he used a brush to transfer pollen from his chosen type of pea plant to fertilise the flowers. All the offspring of the cross between tall and short pea plants were tall.

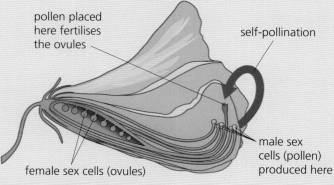

↑ Mendel prevented self-pollination so that he could control which pea plants produced offspring.

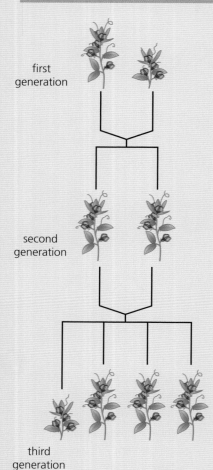

↑ There were no short plants in the second generation, but some appeared in the third generation.

When these tall plants in the second generation fertilised each other, some of the third generation were short.

Mendel's theory

Mendel's theory was that two 'factors' control each feature – one factor from each parent. These 'factors' are now called genes.

The genes in a pair can be the same or different. One gene makes peas tall; another makes them short. When peas have one of each, they are always tall.

Genes that stop other genes having an effect, like the gene for tall pea plants, are called **dominant** genes.

Dominant and recessive genes

So where did the short plants come from in the third generation?

Mendel invented a code to explain what happens. In the diagram **T** shows the dominant gene and **t** shows the version that makes plants short. This is called a **recessive** gene because it does not show up when there is a dominant gene with it.

Variation and classification

In the first generation, the tall plants have two **T** genes, so all their sex cells (pollen and ovules) have one. The short plants have two **t** versions, so all their sex cells have a **t** gene.

The second generation plants inherit a gene from each parent so they have both sorts: **T** and **t**. The **T** version is dominant so the pea plants are all tall.

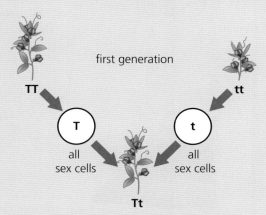

↑ All the 2nd generation are tall because they inherit one of the dominant genes

Each of the second generation plants makes two sorts of sex cells. On average, half carry **T** and half carry **t**, so many of the offspring will have **Tt** again. But male sex cells carrying **T** or **t** could fertilise female sex cells with the same gene. So plants with **TT** or **tt** genes could also be made.

The **TT** and **Tt** plants are tall. But the **tt** plants are short because they don't have the dominant **T** gene. In a cross like this where both parent plants have one of each gene, there is a 1 in 4 chance that their offspring will be short.

Mendel found six other pea features that were inherited in the same way as tallness. Breeding experiments prove that other organisms can inherit different versions of genes in the same way as pea plants.

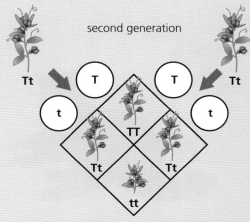

↑ There is a 1 in 4 chance that the peas in the third generation will inherit two recessive genes and be short.

Q

1. Did Mendel's second generation pea plants inherit their parents' features?
2. How did Mendel make his tall and short plants fertilise each other?
3. How can you tell that the gene that makes peas tall is a dominant gene?
4. Why are the second-generation plants all tall?
5. Explain why some of the third-generation plants are short.
6. Draw a diagram to show what would happen if a tall pea plant with **Tt** genes was crossed with a short plant.

- Our features are influenced by the genes we inherit.
- Genes come in pairs and organisms can inherit two different versions of a gene.
- When one gene of a pair is dominant, it controls a characteristic.

17.5 Selective breeding

Objective
- Describe how selective breeding can produce new varieties

Cats

All breeds of cat share one scientific name: *Felis catus*. They are all descended from desert wildcats that lived in the Middle East 10 000 years ago.

How could they end up looking so different?

↑ All these pet cats have inherited different characteristics.

Variation

Wild cats show a lot of variation. Each cat inherits a unique combination of genes from its parents, so they all look different and behave differently.

About 10 000 years ago the tamest ones started to live with humans. The pet cats we have now were produced by **selective breeding** from these cats. There are four steps involved:

1. Decide what characteristics you want the next generation to have.
2. Choose parents that have some of these features and breed them together.
3. Select the offspring with the characteristics you prefer, and breed these together.
4. Repeat the process over many generations.

If you want a small, black cat with big ears, you could mate a male black cat with a small female cat that has big ears. Then keep the kittens that are smallest and blackest and have the biggest ears. These will be your next set of parents. Eventually you could produce a new breed with all three features.

↑ Each kitten inherited some genes from each parent.

Better farm animals

These Holstein-Friesian cows have been selectively bred to give far more milk than their ancestors would. They will give birth, and start producing milk, before they are 2 years old.

Other breeds produce less milk but are better at resisting disease. If they are crossed with Holstein-Friesians, their offspring may inherit useful characteristics from both breeds.

Other breeds of cow have been selectively bred to be strong enough to pull a cart or a plough.

↑ Holstein-Friesian cows produce a lot of milk.

Better crops

This is maize – one of our main cereal crops. Evidence from fossils shows that the plant has been selectively bred for at least 10 000 years. Its wild ancestor produced just a few tiny seeds. Modern maize plants produce much bigger seed heads than their distant ancestors did.

Mutations

Occasionally mistakes are made when cells copy their genes. These are called **mutations**. Most mutations are harmless, but some mutations stop genes working properly. They cause genetic diseases.

Very occasionally, mutations produce new individuals that look different or have characteristics that make them more useful.

Sweetcorn is a variety of maize with a mutated gene. The gene makes it store more sugar and less starch in its seeds, so it tastes sweet.

Plant and animal breeders try to pass the most useful genes in a population to the next generation.

⬆ Modern maize plants produce much bigger seed heads than their distant ancestors did.

⬆ Different genes produce different seed colours.

Q

1. Why do members of a species show a lot of variation?
2. List the steps involved in selective breeding.
3. Explain why selective breeding is useful to farmers.
4. Give an example of an animal that has been selectively bred to be more useful to humans.
5. Describe one way a plant can be made more useful by selective breeding.

- Members of a species vary because they have different combinations of genes.
- Selective breeding produces animals with specific features by controlling which animals produce offspring.
- Over many generations this can produce new varieties of plant and new breeds of animal.

Enquiry 17.6

Developing a theory

Theories

Objective
- Describe how new theories are developed

Why are there so many different living things?

As he struggled to answer this question, Charles Darwin developed one of the most important scientific theories we have – the theory of evolution.

Theories explain how or why things happen. There aren't many scientific theories, because each theory explains a lot of observations and answers a lot of questions.

For example, cell theory states that all living things are made of cells. We use it to explain how living things stay alive, grow, develop, and reproduce.

Most theories form gradually. Scientists collect observations and ask questions. They suggest possible explanations and collect evidence to support their ideas. If the evidence is strong enough, their ideas are accepted by other scientists. Their explanations may eventually link to form a new scientific theory.

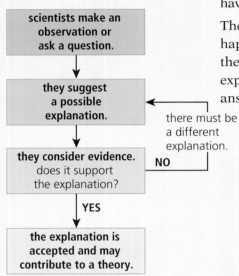

Explanations supported by evidence help scientists produce new theories.

The theory of evolution explains why there are so many different species on Earth.

Making observations and asking questions

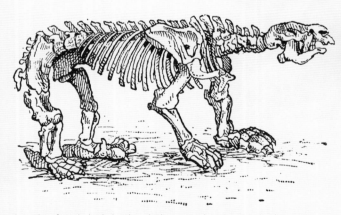

This fossil sloth had a skeleton as big as an elephant.

Modern sloths resemble their ancestors but they are much smaller.

Darwin left university in 1831 and spent 5 years travelling around the world. Most of this time was spent exploring the coast of South America and the nearby Galapagos Islands.

He found fossils of giant mammals. They were similar to modern animals but not identical. One had a skeleton as big as an elephant's, but it was similar to a sloth. Sloths are much smaller. Could the animals have changed over time?

Variation and classification

In the Galapagos Islands, he found different mockingbirds on different islands. Could one type of bird produce four different species?

Darwin also found many different finches. Some had short, strong beaks. They used them to crush nuts and seeds. Others used long, thin beaks to dig for insects. Could these finches share the same ancestors?

↑ Galapagos mockingbirds from the largest island.

↑ Darwin was surprised to find different animals on different islands.

Darwin found different species of mockingbird on the three smaller islands.

A possible explanation

A Scottish geologist called Charles Lyell published evidence that Earth is very old. It showed that rocks change gradually over many years. Darwin thought that living things also changed with time.

A storm blew the first finches to the Galapagos Islands. Over many generations, their beaks changed. Now their beak shapes suit the food each species eats; but how could these changes happen?

Evidence

Darwin collected evidence about selective breeding (see page 216). He realised that some sort of selection always takes place. Most of a plant or animal's offspring don't survive. They compete with each other to survive and reproduce. This is **natural selection**.

Natural selection relies on survival of the fittest. The finches in each group compete for a particular type of food. The finches with the most well adapted beaks get more food, so they are more likely to survive. They pass their useful characteristics to future generations. Over many generations, each group of finches evolves into a different species with a different beak shape.

↑ This stamp celebrates Darwin's achievements.

Stronger evidence

Soon after Darwin published his theory, Mendel published the results of his experiments (see page 214). Unfortunately Darwin never heard about Mendel's work. It would have strengthened his theory by showing how characteristics can be inherited.

Since then many other scientists have added explanations and evidence to Darwin's theory. We now know that genes control inherited characteristics. We also know what genes are made of and how they are passed on to future generations (see page 212).

Q

1. Describe an observation that suggests species change over time.
2. Describe an observation that suggests that one species can split into several different species.
3. Sketch a beak suitable for collecting insects and a beak suitable for crunching seeds and nuts.
4. Explain how a bird's beak shape could affect its survival.
5. Explain the difference between selective breeding and natural selection.

- Developing a theory involves making observations, asking questions, suggesting explanations, and gathering evidence to support them.

17.7 Darwin's theory of evolution

Objective
- Describe Darwin's theory of evolution by natural selection

The theory
Darwin's theory of evolution by natural selection depends on a few simple ideas:

- Living things produce many offspring, but most don't survive.
- Every individual is different and some are more likely to survive than others.
- Succsessful individuals pass their characteristics to their offspring.
- Over time, these characteristics become more common.

Overproduction of offspring
Warthogs like these start reproducing when they are 2 years old. Their numbers could double in size every year if none of their offspring died.

Other animals produce more offspring at a time so their numbers can build up more quickly. One female rabbit can produce up to 40 rabbits in a year.

Why aren't there more of these animals?

↑ Warthogs can double their numbers in a year.

Competition and selection
The number of plants or animals in a population doesn't usually change much. Plants compete for patches of soil, water, and sunlight and animals compete for food and places to live. They both need to avoid being eaten. Most plants and animals die before they have chance to reproduce.

Survival of the fittest
The individuals in a population vary, and some are better adapted to their environment than others. A warthog that could run faster might escape from its predators. It could survive long enough to produce offspring. A cheetah with a better hunting strategy is more likely to catch its prey. It will collect enough food to raise its offspring successfully.

↑ Animals with useful characteristics are more likely to survive.

An animal's predators or prey are part of its environment. Other aspects of their environment can also decide which animals survive, such as the temperature, the water supply, and the presence of disease organisms. This is natural selection.

Inheritance of useful characteristics

Well adapted animals produce more offspring. Their useful characteristics pass to the next generation. Over time, these characteristics become more common and the species changes. It becomes better adapted to its environment. This is evolution by natural selection.

If members of a species become separated, natural selection may turn each group into a different species.

Evolution never stops. As environments change, the characteristics that make species successful change too, so natural selection starts to produce new adaptations.

Resistance

↑ Many rat populations are resistant to poisons that used to kill them.

Since 1950, a chemical called warfarin has been used to be used to poison rats. Like all animals, rats show a lot of variation. Some are not affected by the poison. Whenever the poison is used, it kills most of the normal rats. The resistant ones survive and breed, so they pass the genes that cause their resistance to the next generation. Eventually whole populations of rats can become resistant to warfarin.

Q

1. Plants and animals keep producing offspring, but the total number of each species usually stays the same. Why is this?
2. List some of the things plants and animals compete for.
3. Explain why some animals are more likely to survive than others.
4. Give an example of a characteristic that could improve an animal's chance of surviving for long enough to reproduce.
5. Explain how evolution by natural selection could have made modern cheetahs run faster than their ancestors.

- Members of a species have different characteristics.
- These make some individuals more likely to survive than others.
- The survivors reproduce and pass their useful characteristics to the next generation.
- Over many generations these characteristics become more common and the species changes.

Extension 17.8: Moving genes

Objective
- Describe how living things can be genetically engineered to make new products

Adding genes

Four of these baby mice are not like the others. They glow green under ultraviolet (UV) light.

Scientists added a jellyfish gene to their eggs. They put it into the egg nucleus alongside their other genes. Adding genes to cells like this is called **genetic engineering**.

Genetic engineering is tricky to carry out, but it has advantages over selective breeding. It can be used to add genes that aren't usually found in that species, and it can produce animals with new characteristics more quickly.

↑ The cells in the glowing mice contain a gene from jellyfish.

Copying genes

The engineered mouse egg cells were fertilised with normal sperm, and began to multiply. They kept splitting in two to form all the cells in each embryo. Each new cell had a copy of the glow gene.

Glowing mice can be used for medical research. The glow can show what is happening inside their cells.

Growing medicines

It takes skill and hard work to make new genes work in animals or plants. Bacteria are easier to use. Their cells are simpler. They have no nucleus and their genes are loose in the cytoplasm. They also reproduce very rapidly.

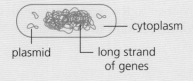

↑ Bacterial cells don't have a nucleus.

Extra genes have been added to many bacteria. One of these is the gene for human insulin. Most people make insulin in their own bodies, but people with diabetes can't. They need to inject insulin into their blood when they need it. They can only remain healthy if they have a regular supply of insulin. The bacteria with the insulin gene produce human insulin for these people to inject.

Variation and classification

Transferring genes to bacteria

Most of the genes in bacteria are in one big strand. The rest are in little loops called **plasmids**. It is easier to insert new genes into these plasmids than into the big strand. The process involves three main steps:

- Identify and extract the gene you want to use.
- Insert the gene into a plasmid.
- Add the plasmid to bacteria.

Bacteria reproduce very quickly in big tanks, and pass copies of their new gene to all their offspring. All the new bacterial cells then make the product from the new gene.

Pharming

Genetically engineered plants and animals can also be used to produce medicines. This is called **pharming**.

When the insulin gene is added to flowering plants, insulin can be extracted from their seeds. But some scientists worry about growing medicines in plants. Their pollen or seeds could contaminate food crops or cause other problems we can't predict.

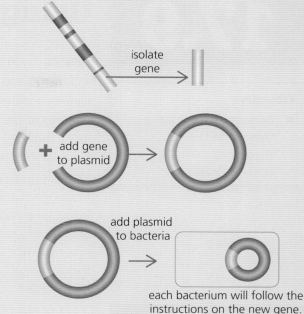

↑ Genetic engineering involves 3 main steps.

The insulin gene has been added to cow and goat embryos. When the animals grow up they will make insulin in their milk. People won't actually drink the milk. The insulin will be separated out. Animals produce a lot of milk all year round, so the price of insulin could drop.

Researchers in Dubai hope to put the insulin gene into camels. They don't produce as much milk as cows but they are more resistant to disease, cheaper to feed, and better adapted to hot, dry climates. If the scientists are successful, life-saving drugs such as insulin could be produced very cheaply in the Middle East and North Africa.

↑ This animal's milk contains a valuable medicine.

Artificial spiders

Scientists think we could get microbes to make nearly everything we need by adding new genes to them. This spider's silk could make ultra-light fabrics with amazing strength and elasticity.

Spiders' silk is far better than manufactured silk, but we can't get enough of it. Spiders tend to eat each other if many are kept together, so scientists put their silk-making genes in bacteria. Silk is toxic to most bacterial cells, but *Salmonella* bacteria can make it and survive. They throw the silk out as fast as they make it.

↑ Spiders' silk is stronger than steel and much lighter.

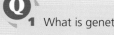

1. What is genetic engineering?
2. Give a reason why genetic engineering is easier in bacteria.
3. Use diagrams to explain how new genes are transferred to bacteria.
4. Explain why genetic engineering can make micro-organisms more useful.
5. Suggest why some people worry about genetically engineered plants but not about genetically engineered bacteria.

- Living things can be genetically engineered to make useful products by adding genes from other organisms to them.

Extension 17.9

Using genes

Objective
- Understand that selective breeding and genetic engineering can be used to produce medicines

Fever

Abha has a fever. She caught malaria when a mosquito bit her. It happens to millions of people every year. At least half a million people die each year because they can't afford medicines.

Malaria is caused by protozoa. Mosquitoes inject them into your blood (see page 41). Malaria victims used to buy medicines to help them recover. They were very cheap. Unfortunately, these medicines no longer work.

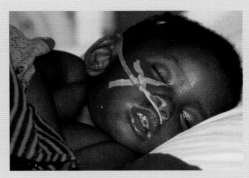
↑ Malaria protozoa infect millions of children every year.

A few of the protozoa had genes that stopped the medicine working. These **resistant** protozoa survived, reproduced, and passed on their genes. Now, most populations of the protozoa are resistant, so these medicines no longer work.

The best cure is an expensive chemical called artemisinin. To stop resistance to this chemical developing, it is mixed with other chemicals to make a medicine called ACT. Microbes rarely develop resistance to more than one chemical at once, so there should always be a cure for people who can afford it.

Chinese herbalists have used artemisinin for thousands of years. They extract it from the leaves of artemisia plants. It costs a lot because it is difficult to extract.

Chemists can make many medicines, but artemisinin is too complicated.

↑ Each leaf contains a small quantity of artemisinin.

Selective breeding

Arteminisin would be cheaper if plants made more of it. UK scientists checked thousands of artemisia plants. The amount of artemisinin in their leaves varied, but a few plants gave a much higher yield than the rest.

By repeatedly breeding the best plants with each other, scientists produced new varieties with even higher yields.

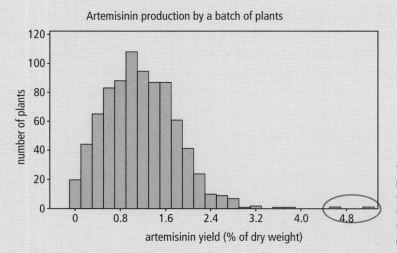

The circled plants have higher yields than the rest, so they will be bred to produce the next generation.

↑ Every population shows variation.

Genetic engineering

US scientists used **genetic engineering** to produce artemisinin more cheaply.

Scientists found the genes that plants use to make the chemical, and put them into yeast cells. This made the yeast cells produce artemisinin.

All living things can read each other's genes. Single-celled organisms reproduce quickly, so they copy their genes quickly and make lots of the proteins these genes code for.

It takes a lot of expertise to make genes work properly in another living thing, but other medicines have been made like this for many years. The genetically engineered micro-organisms are grown in huge tanks, all over the world.

↑ Micro-organisms can be grown in tanks anywhere in the world.

Artemisia plants take more than a year to grow, but yeast cells reproduce quickly. They can produce artemisinin as fast as it is needed. It is also cheaper to extract the chemical from yeast cells than from plant cells.

Scientists hope that all malaria victims will get the best medicine in future.

Stamping out malaria

The companies that make medicines can't cure malaria by themselves. It costs billions to find new medicines and develop ways of producing them. Most malaria victims are too poor to pay high prices for medicines.

Governments and charities all over the world work together to tackle the disease. Billions of dollars have already been spent, but the protozoa that cause malaria are proving difficult to destroy.

Q
1 Explain why most of the medicines used to treat malaria no longer work.
2 Where does artemisinin come from and why is it expensive?
3 How could selective breeding be used to make artemisinin cheaper?
4 Explain how genetic engineering increases the amount of artemesinin produced.
5 Explain why governments and charities need to raise money to tackle malaria.

- Plants and animals produce complicated chemicals that are difficult for chemists to make.
- Genes from plants and animals can make bacteria or yeast cells produce these useful chemicals.
- Micro-organisms grow and reproduce faster than plants and animals so they can produce useful chemicals faster.

Review 17.10

1 A student found these insect larvae A–C in a stream. Use the key to identify them. [4]

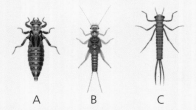

1	Has tail filaments	go to 2
	Has no tail filaments	**dragonfly larva**
2	Has 2 tail filaments	**stonefly larva**
	Has 3 tail filaments	**damselfly larva**

2 These invertebrates are all found in fresh water.

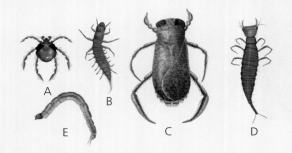

a Use the key below to identify A, C, and E. [3]

b Write a question to distinguish between species B and D. [1]

3 Karis has inherited characteristics from each of her parents.

a Describe one characteristic Karis has inherited from her mother. [1]

b Describe one characteristic she has inherited from her father. [1]

c Explain why she looks a bit like each parent. [2]

4 The diagram shows three human cells.

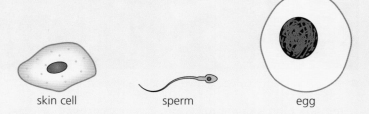

a Name the part of each cell that contains genetic material. [1]

b Describe what happens to a cell's genes when it divides to form two new cells. [1]

c How is the collection of genes in an egg or sperm cell different from those in other cells? [2]

d How are genes from each parent passed to their offspring? [1]

5 There are many different breeds of dog. They are all descended from the same distant ancestors.

Name the technique used to produce these different breeds. [1]

6 A plant breeder wants to produce a new variety of maize. It must have short stems and produce a high mass of grain. None of his existing varieties has both these characteristics.

Variety	Stalk length	Mass of grain
A	short	low
B	long	low
C	long	high

Describe three steps the breeder should take to produce the new variety. [3]

7 Peppered moths rest on tree trunks during the day. Their main predators are birds. The moths come in two forms – dark and light. The light forms are well camouflaged on clean tree trunks.

The percentages of dark and light moths were measured in the years 1700 and 1840.

| Moth colour | % of total population of moths | |
	1700	1840
black	1	90
white	99	10

a Describe how the population of moths changed between 1700 and 1840. [2]

b Between these dates the use of coal increased. Smoke from the burning coal turned tree trunks black. Which type of moth would birds see more clearly on the dark tree trunks? [1]

c Suggest how the change in the colour of tree trunks could change the population of moths. [2]

8 Which of these statements about natural selection are true? [2]

  a Natural selection occurs when some individuals are more suited to their environment.

  b Natural selection causes individuals to evolve.

  c Natural selection explains how species can evolve.

9 Darwin found small birds called finches on each of the Galapagos Islands.

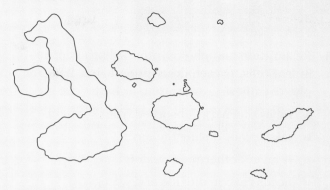

On each island the birds ate different foods and had different beak shapes.

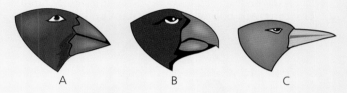

a Which finch has the most useful beak for crushing hard seeds? [1]

b Which finch has the most useful beak for pulling insects through narrow gaps? [1]

c What controls the shape of a finch's beak? [1]

d How could natural selection produce birds with different beak shapes on each island? [2]

e The finches on the Galapagos Islands have different beak shapes, but most of their genes are very similar. Suggest why. [1]

10 Armadillos feed by digging insects out of the soil.

Darwin found fossilised animals in South America. They resembled armadillos but had smaller front claws.

Use Darwin's theory of evolution to explain how armadillos' claws have changed. [2]

Review Stage 9

1 Respiration and photosynthesis both happen in leaf cells. Is each description below true for photosynthesis, respiration, or both?

   a relies on energy absorbed from sunlight [1]

   b takes place during the day [1]

   c is an endothermic reaction [1]

   d needs carbon dioxide and water [1]

   e releases energy from glucose [1]

   f is essential for the survival of the plant [1]

2 The diagram below shows a plant cell.

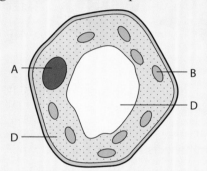

Which letter shows the part that absorbs light energy for photosynthesis? [1]

3 A student grew three identical seedlings at different temperatures.

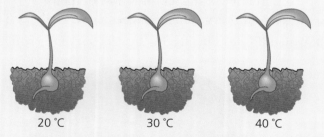

Temperature (°C)	Height of seedling (mm)	
	Day 0	Day 10
20	10	24
30	10	29
40	10	17

   a Which temperature gave the fastest growth? [1]

   b All the seedlings were the same height at the start. List two other variables that should be kept the same in this experiment. [2]

   c Suggest how the results could be made more reliable. [1]

4 The image below shows the reproductive organs in a flower.

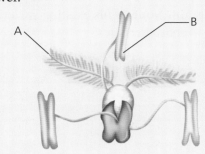

   a Describe how parts A and B help the plant to produce seeds. [2]

   b Explain why this flower does not produce nectar and coloured petals. [2]

5 Pollination and fertilisation are important stages in a plant's life cycle.

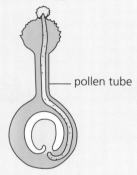

   a Describe what happens during pollination. [1]

   b After fertilisation a pollen tube grows. Explain how the pollen tube aids fertilisation. [2]

Scientists kept pollen grains at a range of different temperatures. They measured the percentage that produced pollen tubes at each temperature.

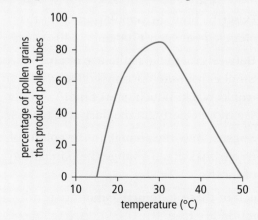

   c What was the optimum temperature for pollen tube growth? [1]

   d Suggest how soya bean yields could be affected if global warming raises summer temperatures. [1]

6. Faizah investigated the effect of competition on seedlings. She sowed different numbers of seeds on damp cotton wool in identical dishes.

20 seeds   40 seeds   80 seeds

All the seeds germinated. The table shows how many of the seeds were still alive after 10 days.

Seeds planted	Seedlings alive after 10 days	% survival after 10 days
20	19	95
40	34	85
80	49	61

a Which variable did Faizah change? [1]

b What conclusion can be drawn from her results? [1]

c Name to things seedlings compete for. [2]

d Plants reduce competition between their seeds by dispersing them. List three ways seeds can be carried away from their parent plant. [3]

7. Jamil investigates mineral deficiency. He compares a plant given every mineral with plants missing magnesium, nitrogen, or phosphorus.

Plant	A	B	C
roots	short	long	short
stem	short, weak	tall, strong	tall, weak
leaves	small, green	large yellow	large, green

a How do plants use magnesium? [1]

b What is nitrogen needed for? [1]

c The plant that received every mineral was tall and had long roots, a strong stem, and large green leaves. Identify the minerals plants A, B, and C are missing. [3]

8. Gila monsters are reptiles. They have thick, waterproof, orange skin. They store fat in their tails and can survive for months without eating or drinking.

Suggest what sort of habitat gila monsters are adapted to live in. [1]

9. Use the key below to identify the groups animals A–C belong to. [3]

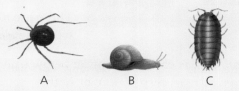

A   B   C

1	Does it have legs?	Yes – go to 2
		No – go to 4
2	Does it have more than three pairs of legs?	Yes – go to 3
		No – **insect**
3	Does it have 2 antennae on is head?	Yes – **crustacean**
		No – **arachnid**
4	Does it have a hard shell?	Yes – **mollusc**
		No – **annelid**

10. Scientists estimated the total biomass of living things in a habitat.

Organism	Biomass (kg)
grass	10 000
grasshoppers	1000
toads	100
snakes	5

a Grass absorbs light energy for photosynthesis. Write the word equation for this reaction. [2]

b Name the primary consumers in this habitat. [1]

c Explain why the total biomass of snakes is so much less than the total biomass of grasshoppers. [2]

d When plants and animals die their bodies rapidly disappear. Explain why. [1]

11. Joy and Naomi are identical twins. Their hair is similar to their mother's but they have their father's chin.

Explain why the twins have some characteristics from each parent. [2]

Reference 1

Choosing apparatus

There are many different types of scientific apparatus. The table below shows what they look like, how to draw them, and what you can use them for.

Apparatus name	What it looks like	Diagram	What you can use it for
test tube			• Heating solids and liquids. • Mixing substances. • Small-scale chemical reactions.
boiling tube			• A boiling tube is a big test tube. You can use it for doing the same things as a test tube.
beaker			• Heating liquids and solutions. • Mixing substances.
conical flask			• Heating liquids and solutions. • Mixing substances.
filter funnel			• To separate solids from liquids, using filter paper.
evaporating dish			• To evaporate a liquid from a solution.
condenser			• To cool a substance in the gas state, so that it condenses to the liquid state.

Reference 1: Choosing apparatus

Apparatus			Use
stand, clamp, and boss			To hold apparatus safely in place.
Bunsen burner			To heat the contents of beakers or test tubes. To heat solids.
tripod			To support apparatus above a Bunsen burner.
gauze			To spread out heat from a Bunsen burner. To support apparatus such as beakers over a Bunsen burner.
pipette			To transfer liquids or solutions from one container to another.
syringe			To transfer liquids and solutions. To measure volumes of liquids or solutions.
spatula			To transfer solids from one container to another.
tongs and test tube holders			To hold hot apparatus, or to hold a test tube in a hot flame.

231

Reference 2

Working accurately and safely

Using measuring apparatus accurately

You need to make accurate measurements in science practicals. You will need to choose the correct measuring instrument, and use it properly.

Measuring cylinder

Measuring cylinders measure volumes of liquids or solutions. A measuring cylinder is better for this job than a beaker because it measures smaller differences in volume.

To measure volume:

1. Place the measuring cylinder on a flat surface
2. Bend down so that your eyes are level with the surface of liquid
3. Use the scale to read the volume. You need to look at the bottom of the curved surface of the liquid. The curved surface is called the **meniscus**.

Measuring cylinders measure volume in cubic centimetres, cm^3, or millilitres, ml. One cm^3 is the same as one ml.

Thermometer

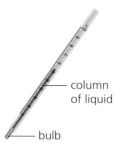

↑ The different parts of a thermometer.

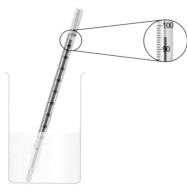

↑ The temperature of the liquid is 95 °C.

These diagrams show an alcohol thermometer. The liquid inside the thermometer expands when the bulb is in a hot liquid and moves up the column. The liquid contracts when the bulb is in a cold liquid.

To measure temperature:

1. Look at the scale on the thermometer. Work out the temperature difference represented by each small division.
2. Place the bulb of the thermometer in the liquid
3. Bend down so that your eyes are level with the liquid in the thermometer
4. Use the scale to read the temperature.

Most thermometers measure temperature in degrees Celsius, °C.

Balance

A **balance** is used to measure mass. Sometimes you need to find the mass of something that you can only measure in a container, like liquid in a beaker. To use a balance to find the mass of a liquid in a beaker:

1. Place the empty beaker on the pan. Read its mass.
2. Pour the liquid into the beaker. Read the new mass.
3. Calculate the mass of the liquid like this:

    (mass of liquid) = (mass of beaker + liquid) − (mass of beaker)

Balances normally measure mass in grams, g, or kilograms, kg.

↑ The balance measures mass.

Reference 2: Working accurately and safely

Working safely

Hazard symbols

Hazards are the possible dangers linked to using substances or doing experiments. Hazardous substances display **hazard symbols**. The table shows some hazard symbols. It also shows how to reduce risks from each hazard.

Hazard symbol	What it means	Reduce risks from this hazard by…
	Corrosive – The substance attacks and destroys living tissue, such as skin and eyes.	• Wearing eye protection • Avoiding contact with the skin
	Irritant – The substance is not corrosive, but will make the skin go red or form blisters.	• Wearing eye protection • Avoiding contact with the skin
	Toxic – Can cause death, for example, if it is swallowed or breathed in.	• Wearing eye protection • Wearing gloves • Wearing a mask, or using the substance in a fume cupboard
	Flammable – Catches fire easily.	• Wearing eye protection • Keeping away from flames and sparks
	Explosive – The substance may explode if it comes into contact with a flame or heat.	• Wearing eye protection • Keeping away from flames and sparks
	Dangerous to the environment – The substance may pollute the environment.	• Taking care with disposal

Other hazards

The table does not list all the hazards of doing practical work in science. You need to follow the guidance below to work safely. Always follow your teacher's safety advice, too.

- Take care not to touch hot apparatus, even if it does not look hot.
- Take care not to break glass apparatus – leave it in a safe place on the table, where it cannot roll off.
- Support apparatus safely. For example, you might need to weigh down a clamp stand if you are hanging heavy loads from the clamp.
- If you are using an electrical circuit, switch it off before making any change to the circuit.
- Remember that wires may get hot, even with a low voltage.
- Never connect wires across the terminals of a battery.
- Do not look directly at the Sun, or at a laser beam.
- **Wear eye protection** – *whatever* you are doing in the laboratory!

Reference 3

Recording results

A simple table

Results are easier to understand if they are in a clear table.

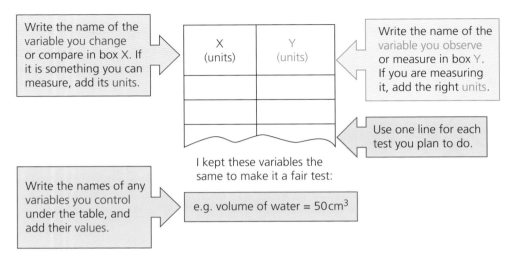

Write the name of the variable you change or compare in box X. If it is something you can measure, add its units.

Write the name of the variable you observe or measure in box Y. If you are measuring it, add the right units.

Use one line for each test you plan to do.

Write the names of any variables you control under the table, and add their values.

I kept these variables the same to make it a fair test:

e.g. volume of water = 50 cm³

Units

It is very important to use the correct units.

    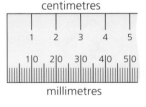

The units of **temperature** include **degrees celsius (°C)**.

The units of **time** include **seconds (s)**, **minutes (min)**, and **hours (hr)**.

The units of **volume** include **cubic centimetres (cm³)**, and **cubic decimetres (dm³)**.

The units of **mass** include **grams (g)**, **kilograms (kg)**, and **tonnes (t)**.

The units of **length** include **millimetres (mm)**, **centimetres (cm)**, **metres (m)**, and **kilometres (km)**.

Reference 3: Recording results

Making results reliable, and reducing error

You should always try to repeat observations or measurements. Never rely on a single result. If repeat results are similar, they are more **reliable**. Repeating results also helps to reduce error. If the results keep changing, try to find out why. It could be a mistake, or there might be another variable you need to control.

When you collect similar measurements, you should calculate their average value.

Three students find the time it takes to draw a table. Jamil takes 75 seconds, Abiola takes 35 seconds and Karis takes 73 seconds.

Abiola's result is **anomalous** because it is very different from the others. Jamil and Karis find out why. Abiola's table is very messy. She did not use a ruler. They decide to leave it out of the average.

$$\text{average time} = \frac{\text{sum of the measurements}}{\text{number of measurements}} = \frac{75 + 73}{2} = \frac{148}{2} = 74 \text{ seconds}$$

A table for repeat results

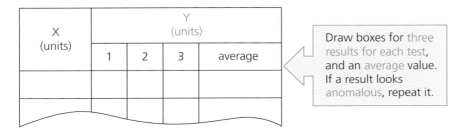

Draw boxes for three results for each test, and an average value. If a result looks anomalous, repeat it.

A table for results that need to be calculated

Some variables can't be measured. They need to be calculated from two different results. Do all the calculations before you find the average value.

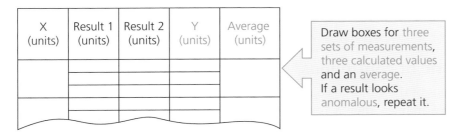

Draw boxes for three sets of measurements, three calculated values and an average. If a result looks anomalous, repeat it.

Reference 4

Displaying results

Drawing a bar chart

Three students timed how long they spent doing homework. The results are in the table.

A bar chart makes results like these easier to compare.

Students	Time spent on homework (hours)
Deepak	2
Jamila	4.5
Kasim	1

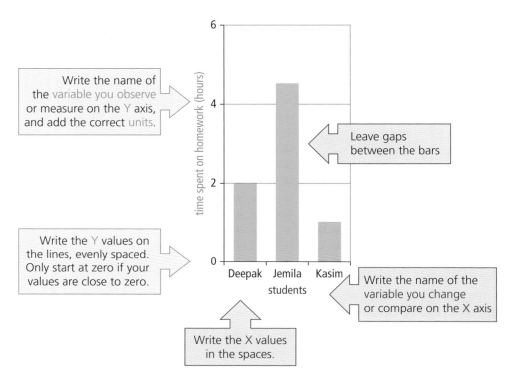

Categoric or continuous

If the values of the variable you change (X) are words or names, then X is a **categoric** or **discrete** variable. You can only draw a bar chart for this type of variable.

Variables like shoe size are also categoric variables. They are numbers, but there are no inbetween sizes.

Other variables are **continuous** variables. Their values can be any number. Height is a continuous variable and so is temperature.

If the variables you change and measure are both continuous variables, display the results on a line graph or scatter plot.

Reference 4: Displaying results

Drawing a line graph

A line graph makes it easier to see the **relationship** between two continuous variables – the variable you change or compare and the variable you observe or measure.

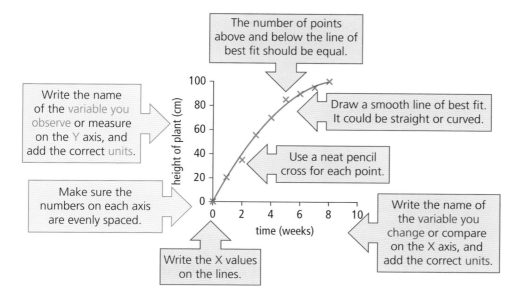

Drawing a scatter graph

A scatter graph shows whether there is a **correlation** between two continuous variables. In the graph below, all the points lie close to a straight line. That means there is a correlation between them.

A correlation does not mean that one variable affects the other one. Something else could make them both increase or decrease at the same time.

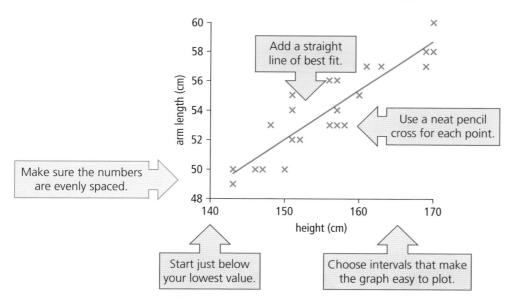

237

Reference 5

Analysing results: charts and diagrams

Describe the pattern.
(e.g. A is bigger than B)

↓

Use the numbers to compare results.
(e.g. A is three times bigger than B)

↓

Suggest a reason using scientific knowledge.
(e.g. This is because...)

Charts and diagrams help you to analyse results. They show differences between results clearly. The flow chart shows you how to analyse results.

Here is a bar chart to show how long it takes three students to get to school.

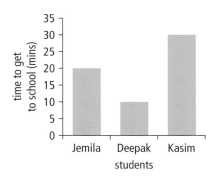

The first stage of analysing your results is to **describe the pattern**.

> "Deepak takes the shortest time to get to school.
> Kasim takes the longest time to get to school."

The next stage it to **use the numbers** on the y-axis of the bar chart to **make comparisons**.

> "Deepak takes 10 minutes to get to school, but Kasim takes 30 minutes.
> Kasim takes three times longer."

Finally, **suggest reasons** for any differences that you have found.

> "This could be because Kasim lives further away or because Deepak walks more quickly."

When you do an experiment, use **scientific knowledge** to explain differences between results. Here are some examples of scientific reasons.

I think that the powdered sugar dissolved faster than the normal sugar because the pieces were much smaller.

I think that the plant near the window grew more quickly than the plant away from the window because there was more sunlight there.

I think that the shoe on the carpet needed a bigger force to move it than the shoe on the wooden floor because there was more friction between the shoe and the carpet.

Reference 5: Analysing results: charts and diagrams

Analysing results – line graphs

Line graphs show correlations between continuous variables. When you have plotted the points on a line graph, draw a line of best fit. Then analyse the graph. The flow chart on the right shows how to do this.

In the graphs below the line of best fit is shown, but not the points.

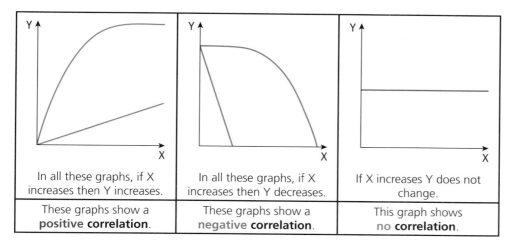

In all these graphs, if X increases then Y increases.	In all these graphs, if X increases then Y decreases.	If X increases Y does not change.
These graphs show a **positive correlation**.	These graphs show a **negative correlation**.	This graph shows **no correlation**.

On some graphs the line of best fit is a straight line. You can say 'Y changes by the same amount for each increase in X'. The blue line in Graph 1 shows that Y increases by the same amount for each change in X. The red line shows that Y decreases by the same amount for each change in X. You can choose values to illustrate this. Graph 2 on the right shows how.

A graph with a positive correlation where the line of best fit is a straight line that *starts at zero* is a special case. In this case Y is **directly proportional** to X. You can say 'if X doubles then Y will also double'. (see Graph 3.) You can choose values to illustrate this.

A graph that has a curved shape with a negative correlation may show that Y is **inversely proportional** to X. Choose pairs of values. If you get the *same number* every time you multiply X by Y then Y is inversely proportional to X. This is the same as saying 'if you double X then Y will halve'. (see Graph 4.) You can choose values to illustrate this.

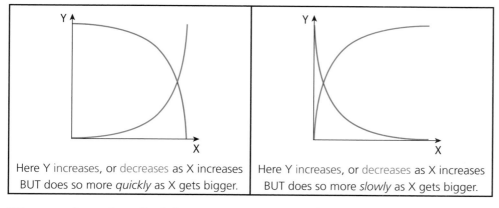

Here Y increases, or decreases as X increases BUT does so more *quickly* as X gets bigger.	Here Y increases, or decreases as X increases BUT does so more *slowly* as X gets bigger.

When you have described the pattern in your results, try to use scientific knowledge to explain the pattern.

Describe the pattern by saying what happens to Y as X increases. (e.g. As X increases Y increases)

↓

Choose pairs of values to illustrate the pattern and compare them. (e.g. When X is 3, Y is 2, and when X is 6, Y is 4 so doubling X will double Y)

↓

Suggest a reason using scientific knowledge. (e.g. This is because...)

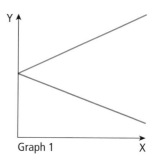

Graph 1

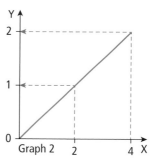

Graph 2

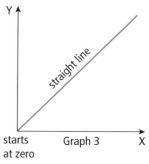

Graph 3

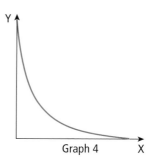

Graph 4

239

Reference 6: Detecting gases

Hydrogen gas

To find out whether a reaction in a test tube has produced hydrogen gas:

1. Place your thumb over the end of the test tube to collect a small quantity of gas.
2. Place a lighted splint in the gas.
3. If the splint goes out with a squeaky pop, hydrogen gas is present.

Oxygen gas

To find out whether a reaction in a test tube has produced oxygen gas:

1. Place your thumb over the end of the test tube to collect a small quantity of gas.
2. Place a glowing splint in the gas.
3. If the splint relights, oxygen gas is present.

Carbon dioxide gas

To find out whether a reaction in a test tube has produced carbon dioxide gas:

- Bubble the gas through limewater solution.
- If the limewater solution turns milky or cloudy, carbon dioxide is present.

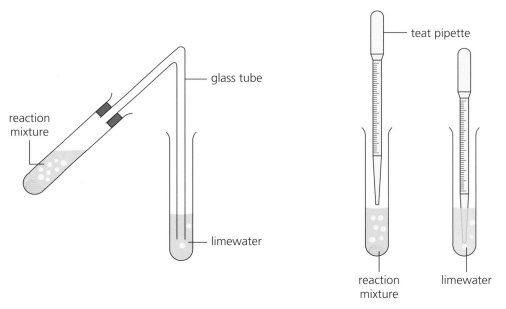

⬆ The diagrams show two ways of bubbling gases through limewater.

Reference 7
Animal classification

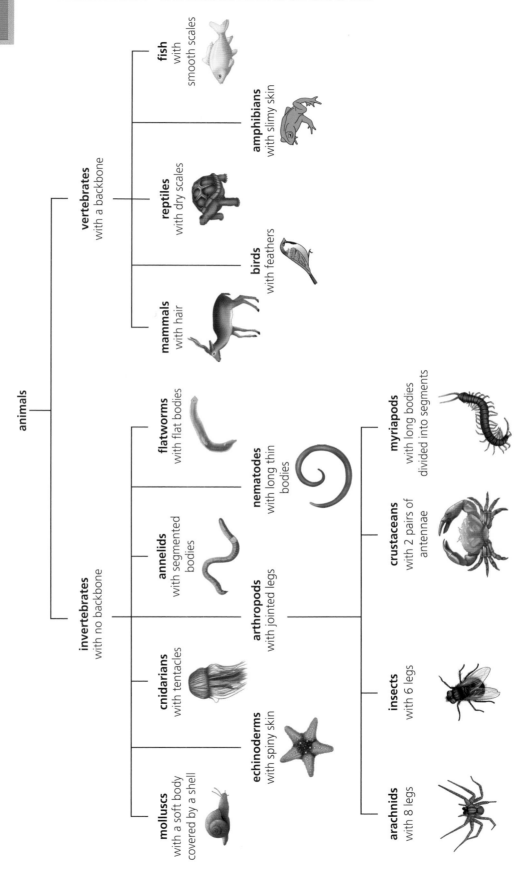

Glossary

Acid rain Rain with a pH of less than 7, formed when an acidic gas dissolves in rainwater.
Active site The part of an enzyme that attaches to molecules and speeds up their reactions.
Active transport The use of energy to move substances across cell membranes.
Adaptation A physical feature or behaviour that suits plants and animals to their environment.
Addictive Describes a drug that changes brain cells so people need to keep taking it.
Aerobic respiration A reaction that uses oxygen to release energy from glucose.
Aeroponics Growing plants with their roots in air instead of soil.
Alcohol The substance produced when yeast ferments sugary solutions.
Algae Organisms found in water. They carry out photosynthesis but lack stems, roots, and leaves.
Alimentary canal The tube that carries food from the mouth to the rectum.
Alveoli Tiny air sacs in the lungs where gas exchange takes place.
Amino acids The type of molecule that makes up proteins.
Anaemia A deficiency disease in humans caused by a lack of iron.
Anaerobic respiration A reaction that produces energy without using oxygen.
Antagonistic Describes muscles that pull in opposite directions.
Anther The upper part of a stamen where pollen is produced.
Antibody A molecule made by white blood cells that helps to destroy pathogens.
Anus The end of the alimentary canal through which faeces exit the body.
Artery A thick-walled blood vessel that carries blood away from the heart.
Arthritis A painful disease that occurs when the cartilage at the ends of bones wears away.
Arthropods The most common type of invertebrate. They all have jointed legs.
Asthma A disease that makes breathing difficult because the bronchioles become too narrow.
Athlete's foot A fungal infection that occurs between the toes and causes the skin to become flaky.
Backbone A structure made up of vertebrae that runs from the base of the skull to the pelvis.
Ball-and-socket joint Joints that allow arms and legs to move in every direction.
Beetles Insects that have tough covers over their wings.
Beri-beri A deficiency disease in humans caused by a lack of vitamin B1.
Bile A substance released into the small intestine to emulsify fats.
Bioaccumulation The gradual build-up of persistent organic pollutants in food chains.
Biodiesel A biofuel made from plant oils.
Biodiversity The number and variety of species present.
Bioethanol A biofuel made out of sugars.
Biological catalysts Naturally occurring molecules (enzymes) that speed up chemical reactions.
Biomass The mass of living material in an area.
Biometrics Identifying people using unique characteristics such as fingerprints.
Brain The organ that takes in data from sense organs, takes decisions, and controls your actions.
Bronchiole Tiny tubes that carry air to the alveoli in lungs.
Bronchus (plural: bronchi) A tube that caries air from the trachea to one lung.
Budding A method of reproduction. Offspring develop as buds on the sides of parent cells.
Caffeine The drug found in coffee which is a stimulant.
Camouflage A colour or pattern that helps an organism to blend in with its surroundings.
Cancer A disease caused by cells dividing faster than they should.
Capillary A very thin-walled blood vessel used to deliver substances to cells.
Captive breeding programme Breeding endangered animals in protected environments such as zoos.
Carbohydrase The type of enzyme that breaks down carbohydrates.
Carbon monoxide A toxic gas that reduces the amount of oxygen red blood cells can carry.
Carbon-neutral Describes fuels that don't change the amount of carbon dioxide in the air.
Carcinogenic Describes a substance that can cause cancer by making cells divide faster than they should.
Carnivore A meat-eating animal.
Carpel The female part of a flower.
Cartilage A smooth tissue that covers the ends of bones and lets them slide over each other.
Catalysts A substance that speeds up chemical reactions.
Category A group that share characteristics.
Cell The building block that makes up living things.
Cell membrane The outer layer of a cell which controls what enters and leaves it.
Cell wall The tough outer covering of plant cells which helps support them.
CFC The type of compound used in early refrigerators which depleted the ozone layer.
Chemical digestion Breaking down large molecules into small molecules using enzymes.

Glossary

Chlamydia A common sexually transmitted infection that can cause infertility.

Chloroplast A green structure found inside plant cells where photosynthesis occurs.

Chromosome Strings of genes made from DNA found in the nucleus of every cell.

Ciliated cell A cell with tiny hair-like structures that can sweep material along tubes.

Circulatory system The heart and blood vessels that transport blood around the body.

Classification Sorting organisms into groups based on their similarities and differences.

Climate change A change in the average weather conditions.

Cold A common infection caused by a virus which causes a runny nose, sore throat, and fever.

Community A group of species that live in the same place.

Conception The creation of life by the fertilisation of an egg cell by a male sex cell.

Condoms Devices that can protect people from sexually transmitted infections.

Conservation Saving living things from extinction and protecting their natural environment.

Consumer An organism that obtains food by eating other organisms.

Continuous variation Variation shown by characteristics that can take any value within a range like height.

Contract Get shorter.

Controlled The value of these variables must not change during an investigation.

Correlation A relationship between two variables.

Cytoplasm The gel-like fluid in cells where chemical reactions take place.

Decomposers Organisms that break down dead plants and animals and their waste products.

Deficiency diseases Diseases that occur when people don't get enough of an essential nutrient.

Deforestation The permanent removal of forests.

Denatured Describes an enzyme when its shape has changed so that it no longer works.

Depressant Type of drug that slows your reactions.

DNA The giant molecule that genes are made from.

Diaphragm A thick sheet of muscle below your lungs that contracts when you breathe in.

Differentiated Describes cells that have become specialised to do specific jobs.

Digestion The process that uses enzymes to break down the large molecules in food.

Digestive system The organs that digest and absorb food.

Discontinuous variation Variation shown by characteristics that can only take set values, such as blood groups.

Dislocate A bone moving out of its socket.

Dominant Describes a gene that always affects an organism even when only one copy is present.

Dormant Alive but not growing, for example a dry seed.

Dough A thick mixture of flour and liquid, used to make bread.

Ecosystem The living things in an area and the environment they interact with.

Egg cell A female sex cell (gamete).

Ejaculate The ejection of semen from a man's penis.

Electron microscope A microscope that uses an electron beam to produce a highly magnified image.

Embryo A new organism that is growing from a fertilised egg but does not have a full set of organs yet.

Emulsifies Breaks down into smaller droplets.

Endothermic Describes a reaction that takes in energy.

Environmental variation Variation between individuals caused by the environmental conditions they experience.

Enzyme Protein that speeds up reactions such as those that break down large food molecules.

Eutrophication The effect of adding too many minerals to a lake or river.

Evolution The way a species gradually changes as a result of survival of the fittest.

Evidence A collection of observations or measurements which may support an explanation.

Excretion The removal of waste products from the body.

Exothermic Describes a reaction that releases energy.

Explanation A possible answer to a question.

Faeces Solid waste excreted from your digestive system.

Ferment Break down sugar into carbon dioxide and alcohol.

Fertilisation Fusing the nucleus of a male sex cell with an egg cell to create a new life.

Fertilised Describes an egg when a sperm cell has fused with it.

Fertiliser A product used to improve plant growth by supplying extra minerals to the soil.

Fetal alcohol syndrome Birth defects caused by a mother drinking too much alcohol while pregnant.

Fibre A food component that cannot be digested but keeps the gut healthy.

Filament The lower part of a stamen in a flower, which supports the anther.

Flaccid Describes plant cells that have lost water and cannot support themselves.

Flowers These contain the reproductive organs of flowering plants. They produce seeds.

Flowering plants Plants that produce flowers and reproduce using seeds.

Flu A common infection caused by a virus which causes a runny nose and tiredness.
Flue gas desulfurisation A method used to remove sulfur dioxide from waste gases.
Food poisoning Vomiting and diarrhoea caused by eating food contaminated with microbes.
Food web Diagrams that show the connections between different food chains.
Fortified Describes food that has extra nutrients added to it.
Frequency chart Chart used to display discontinuous variation by showing the number in each category.
Fruit The structure that develops around a seed.
Fungus A type of micro-organism that absorbs nutrients from its surroundings.
Gall bladder The digestive organ that produces bile.
Gametes Male and female sex cells.
Gas exchange The process that takes place in your lungs when blood swaps carbon dioxide for oxygen.
Genes You inherit these from your parents and they influence the way your body develops.
Genetic engineering Adding new genes to cells.
Germination This happens when a seed starts to grow.
Global warming A gradual rise in the temperature of Earth's atmosphere and oceans.
Glucose A small carbohydrate molecule made by photosynthesis and broken down by respiration.
Gonorrhoea A common sexually transmitted infection that can cause infertility.
Greenhouse A structure that lets sunlight in but stops heat escaping.
Greenhouse effect The way some gases absorb heat and raises the temperature of Earth's atmosphere.
Growth factors Chemical signals released by cells that affect the way other cells develop.

Gut The tube that carries food from the mouth to the rectum.
Habitat The place where a plant or animal lives.
Haemoglobin A chemical that carries oxygen inside red blood cells.
Hallucinogen A type of drug that distorts your senses.
Heart The organ that pumps blood around your circulatory system.
Hepatitis C A serious disease caused by a virus that infects the liver.
Herbicide A substance that stops unwanted plants growing.
Herbivore A plant-eating animal.
Hinge joint The type of joint that allow arms and legs to bend and straighten.
Hormone A chemical messenger that affects cells all over the body.
Hybrid The offspring of two different species. They are usually infertile.
Hydroponics Growing plants with their roots in water instead of soil.
Hyperaccumulators Plants that can absorb large quantities of minerals.
Hyphae The thin threads that make up the body of a fungus.
Ice cores Columns of ice drilled from ice sheets.
Immune system The body system that identifies and destroys most pathogens.
Immune Resistant to a disease.
Immunised Made immune to a disease using a vaccine.
Implantation This takes place when an embryo settles into the wall of the uterus.
Inbreeding When animals breed with close relatives.
Infectious disease Disease caused by micro-organisms that can spread from person to person.
Infertile Describes an organism that is not able to reproduce.
Infertility Inability to reproduce.

Inherited variation Variation caused by inheriting different combinations of genes.
Insecticide A substance that kills insects.
Insects The most common type of arthropod. They all have six legs.
Insulator A substance such as fat or air that minimises heat loss by conduction.
Intercostal muscles Muscles between your ribs that help to create breathing movements.
Interdependent Describes populations that affect each other, such as predators and their prey.
Invasive species A species that outcompetes the existing plants or animals in a habitat.
Iodine A chemical used to detect starch. It turns starch dark blue.
Joint The place where two bones meet.
Kidneys Paired organs in the abdomen that clean blood and produce urine.
Kwashiorkor A deficiency disease caused by a lack of protein in the diet.
Lactic acid A molecule produced when bacteria respire using the sugars in milk.
Ladybirds A type of beetle with brightly coloured, rounded bodies.
Large intestine The organ that absorbs water to turn a mixture of fibre and bacteria into faeces.
Latin The language used to give each species a unique, two-part name.
Leaves The organs in a plant which absorb solar energy and make food.
Leishmaniasis A disease caused by protozoa carried by sandflies.
Lichens Living things made from fungi and algae that can make their own food.
Ligaments The bands of tissue that hold bones together at joints.
Line of best fit A line that comes closest to all of the points on a graph.

Glossary

Living indicators Species that can be used to show how much pollution is present.

Magnify Make something appear bigger than it really is.

Malaria A disease carried by mosquitoes that causes fever and chills.

Malnutrition The result of an unbalanced diet.

Mark and recapture A method used to estimate the number of animals in a population.

Mechanical digestion Using teeth to break food into smaller pieces.

Menopause The time in a woman's life when her periods stop.

Menstrual cycle A repetitive sequence of events that prepares the womb to receive a fertilised egg.

Micro-organism Organisms that are too small to be seen without a microscope.

Microscope An instrument used to make magnified images.

Mitochondria The compartments inside cells where aerobic respiration takes place.

Model A simple way of representing more complicated structures or ideas.

Mucus A sticky fluid that traps microbes so that they can be swept out of your lungs.

Muscle The tissue that contracts and pulls on bones to produce movement.

Muscle fibre A single muscle cell.

Muscular system All the muscles that move bones.

Mutation A change in a gene.

Natural selection The way a population's environment selects which individuals survive.

Negative correlation A relationship in which one variable increases, while the other decreases.

Nervous system The brain, nerves, and sense organs that control your thoughts and actions.

Neutralise React with an acid and convert it to a neutral salt and water.

Nicotine An addictive drug found in cigarette smoke, that acts as a stimulant.

Night blindness A deficiency disease caused by a lack of vitamin A.

Non-renewable An energy resource that will run out and cannot be replaced.

Nucleus The part of a cell that contains genes and controls the cell's activity.

Nutrient A useful substance obtained from food which is needed in a balanced diet.

Obese Someone is obese if their body mass is much higher than average.

Optimum pH The pH at which enzymes work best.

Organic matter Material that came from living things and can be decomposed by microbes.

Organism A living thing.

Organ A collection of tissues that work together, e.g. the heart.

Ovary A female reproductive organ that produces sex cells.

Oviduct One of the tubes in females that carry egg cells from the ovary to the uterus.

Ovules These produce female sex cells inside the ovary of a flowering plant.

Oxides of nitrogen Acidic gases formed when petrol burns in a car engine.

Ozone A gas in the upper atmosphere that stops harmful ultraviolet radiation reaching Earth.

Palisade cell A plant cell found near the top of a leaf. Most photosynthesis takes place here.

Pancreas A digestive organ that releases enzymes into the small intestine.

Pasteurisation Heating food or drink to a temperature that kills microbes without ruining its flavour.

Pathogen A micro-organism that causes disease.

Peak expiratory flow (PEF) The maximum speed at which someone can expel air from their lungs.

Penis The male organ that deposits sperm in the vagina during sex.

Period The bleeding that occurs once a month in females.

Peristalsis The squeezing action the gut walls use to push food along.

Phagocyte A white blood cell that detects and destroys pathogens.

Pitfall trap A container buried in the ground used to trap invertebrates.

Pharming Using genetically engineered plants and animals to produce medicines.

Phloem A tube that carries sugars from leaves to other parts of a plant.

Photobioreactors Tubes surrounded by artificial light in which algae can be grown.

Photosynthesis The reaction green plants use to make their own food from carbon dioxide and water.

Phytoextraction Using plants to remove minerals from the soil.

Plasmid A small loop of genes found in bacteria.

Platelets Small fragments of cells that help blood to clot when a blood vessel is damaged.

Pollination Moving pollen from an anther to a stigma.

Pollution A harmful substance that has been added to the environment, e.g. fertiliser in rivers.

Population The number of individuals in a group of living things.

Positive correlation A relationship in which one variable increases, as the other increases.

Predator An animal that hunts other animals for food.

Prediction The expected outcome of an experiment based on scientific ideas.

Prey An animal that is hunted or killed for food.

Primary consumers Animals that feed on plants or other organisms which make their own food.
Producer An organism that can make its own food like a plant.
Protease The type of enzyme that breaks down proteins.
Protozoa A group of micro-organisms.
Puberty The time when young people's bodies change to prepare them for reproduction.
Pulse The regular throb of an artery that indicates your heart rate.
Quadrat A square frame used to find the number of plants growing in a known area.
Questions Things a scientist wants to find out.
Range The difference between the highest and lowest values in a sample.
Rectum The final part of the large intestine where faeces are stored.
Recessive Describes a gene that has no effect when only one copy is present.
Red blood cell Cells in the blood that carry oxygen around the body.
Rejected A transplanted organ or tissue that is not accepted by the recipient's body.
Renewable Describes an energy resource that will not run out because it is constantly replaced, e.g. wind power.
Repeatable Describes measurements that are the same every time they are taken.
Reproduction The production of offspring.
Resistant Describes individuals that survive in the presence of chemicals that are usually toxic.
Respiration The release of energy from glucose and oxygen in living things.
Respiratory system The organ system containing the lungs and the muscles that control them.
Ribs Curved bones that form a cage around the heart and lungs.

Rickets A deficiency disease caused by a lack of vitamin D or calcium.
Ringworm An infectious disease caused by a fungus growing in your skin.
Root hair cell Cells with long, thin extensions that increase the surface area of roots.
Roots Plant organs that hold the plant in the ground and absorb water and minerals.
Saliva A slippery fluid produced in the mouth which contains a carbohydrase enzyme.
Scaffold A support that gives soft tissue structure and makes it the right shape.
Scavenger An animal that feeds on dead plant or animal material.
Scientific question A question that involves one variable that can be changed and one that can be measured.
Scurvy A deficiency disease caused by a lack of vitamin C.
Secondary consumers Carnivores that feed on primary consumers.
Seed A structure made to protect the embryo when a flowering plant reproduces.
Selective breeding Choosing the parents of the next generation so they inherit specific characteristics.
Semen A fluid containing sperm.
Sense organs These detect changes in your environment and send nerve impulses to your brain.
Sensitivity The ability to detect changes in the environment.
Sewage Dirty water containing urine and faeces.
Sex hormone A chemical that produces the changes that take place during puberty.
Sexually transmitted infections (STIs) Diseases passed from one person to another through sex.
Skeleton The framework of bones and cartilage that supports an organism.

Skull The bone that surrounds and protects a vertebrate's brain.
Sleeping sickness A disease caused by protozoa carried by tsetse flies.
Small intestine The part of the digestive system where most nutrients are absorbed into the blood.
Smallpox An infectious disease that was wiped out using vaccines.
Solar cells Devices that transfer energy from light to electricity.
Specialised Suited to do a specific job.
Species A group of organisms with similar characteristics that can produce fertile offspring.
Specimen A section of tissue that can be observed under a microscope.
Sperm cell A male sex cell.
Sperm duct One of the tubes that carry sperm from the testes towards the penis.
Spinal cord The bundle of nerves inside your backbone.
Spongy mesophyll cells Plant cells near the bottom of leaves with air large spaces between them.
Spontaneous generation A theory that microbes are formed from decaying matter; the theory has been disproved.
Spores Reproductive cells made by moss, ferns, and fungi.
Sprain An injury to a joint caused by overstretching a ligament.
Stamens The male parts of flowers.
Starch A large carbohydrate molecule made from lots of glucose molecules joined together.
Stem cells Cells from an early embryo that can divide to produce any other sort of cell.
Stem The plant organ that holds up the leaves and transports food, water, and minerals.
Sterilise Destroy the micro-organisms in or on something.
Stigma The upper part of a carpel which catches pollen.

Glossary

Stimulant A drug that make you feel alert or speeds up your reactions.

Stoma (plural: stomata) A gap on the underside of a leaf which allows gases to pass in and out.

Sulfur An element found in small amounts in fuel, which burns to create an acidic gas.

Sulfur dioxide An acidic gas formed when most fuels burn.

Survival of the fittest The survival of individuals that are best adapted to their environment.

Style The middle part of a carpel in a flower, which connects the stigma to the ovary.

Streamlined Describes a shape that is narrow at both ends to reduce friction.

Surface area The number of square metres covered.

Synovial fluid A slippery liquid found in most joints which provides lubrication.

Systems Groups of organs that work together to do a specific job.

Tar A mixture of chemicals in cigarette smoke that can paralyse cilia and cause cancer.

Tendon The strong tissue that connects muscle to bone.

Tertiary consumers Animals that feed on secondary consumers.

Tissue A group of similar cells that work together to perform a particular function.

Top predator The animal at the top of the food chain, that is not eaten by other animals.

Trachea The tube that carries air from your mouth and nose to your bronchi.

Transpiration The flow of water through a plant.

Transplant Move tissues or organs from one person to another.

Tree rings The rings of wood in a tree trunk.

Trophic level An organism's position in a food chain. Producers occupy trophic level 1.

Turgid Describes a firm, rigid cell full of water. Its vacuole pushes against the cell wall.

Typhoid An infectious disease spread by eating food or drink contaminated by bacteria.

Ultraviolet radiation A type of light that can damage plants and cause skin cancer in humans.

Urine Fluid excreted by the body that contains excess water and waste products.

Uterus The female organ where a baby develops.

Vaccine A weakened pathogen that trains your immune system to make antibodies to it.

Vacuole A fluid-filled sac in plant cells that helps the cell wall to provide support.

Vagina A tube from the outside of a female's body to the uterus.

Valid A conclusion that can be trusted because it is based on reliable results from fair tests.

Variable Something that can have more than one value.

Variation The differences between individuals.

Vector An organism that carries a pathogen from one person to another.

Vein A thin-walled blood vessel with valves that leads blood towards the heart.

Vertebrae Small bones that make up the backbone which surrounds and protects the spinal cord.

White blood cell Blood cells that can identify or attack pathogens.

Wildlife sanctuaries Special areas where animals can be protected.

Wilt What a plant does when its leaves and stems droop due to lack of water.

Xylem A tube that carries water and minerals from a plant's roots to its leaves.

Yeasts Microscopic single-celled fungi that do not have hyphae.

Index

acid rain 58, 67, 198–9
adaptations 44, 52, 66, 172–3, 182
   extreme habitats 174–5
addiction 131
adolescence 142–3
aeroponics 163
AIDS 151
air pollution
   acidic gases 198–9
   indicators 198
   reducing pollution 199
alcohol 32, 146, 154–6
   fetal alcohol syndrome 147
algae 30, 79
   algal oil 65
alimentary canal 104
altitudes 127
alveoli 125
amino acids 92
amniotic fluid 138
amphibians 76
anaemia 99, 114–15
   sickle cell anaemia 113
anaerobic respiration 128–9
animals 28
   automatic cameras 181
   conservation 60–1, 66, 195
   counting animals 178
   electronic tags 181
   estimating animal numbers 178, 183
   following tracks 180–1
   habitats 52–3, 174–5
   movement 173
   naked mole rats 172
   observing animals 180
   ocean floor 175
   predators and prey 54, 61, 172–3, 183, 191
   reducing heat loss 175
   seed dispersal 169
   selective breeding 216, 226
   survival 176–7
Antarctica 53, 58
anthers 166
antibiotics 153
antibodies 151, 152
anus 105
Arctic 174
Arrhenius, Svante 200
arsenic 165
artemisinin 224–5
arteries 20, 116

arthritis 15
arthropods 75
asthma 132–3
athlete's foot 40

babies 139, 140–1
   premature babies 141
backbone 14
bacteria 31, 32–3, 40
   disease 152, 153
   transferring genes to bacteria 223
   yoghurt and fermented milk drinks 36, 50
beetles 75
beri-beri 98
bile 107
bilharzia 148
bioaccumulation 203
biodiesel 64
biodiversity 188, 191
   comparing biodiversity 194
   conserving animals 195
   importance of forests 194
   saving seeds 195
bioethanol 64, 65
biofuels 62–3, 64–5
biomass 54, 84, 187, 197, 229
biometrics 69
birds 76
   Guam 192–3
blood 20–1, 68, 122–3
   blood pressure 117, 120
   blood tests 113
   red blood cells 44, 112
   white blood cells 112, 150
Bolt, Usain 128, 129
bone cells 45
brain 46
bread 36
bronchi and bronchioles 124–5

caffeine 146
camouflage 172
cancer 130
cannabis 147
capillaries 20, 112, 116
carbohydrases 106
carbohydrates 92, 97
carbon dioxide 59, 64
   decomposers 189
   global warming 200–1
carbon monoxide 131
carcinogens 130
carnivores 54, 186
carpels 166
cartilage 15
catalysts 106
cells 42, 51, 82

cell membrane 42
chromosomes 212–13
eggs 136, 137
palisade cells 85
replacing your cells 49
sensory cells 47
specialised cells 44–5
sperm 136, 137
stem cells 49
tissues 48
CFCs 58
cheese 108
Chernobyl 164
chlamydia 149
chloroplasts 43, 85, 158–9
chocolate 37
chromosomes 212
   genes and chromosomes 212
   sex chromosomes 213
cilia 130
circulatory system 18, 20–1, 116–17, 122–3, 156–7
   blocked tubes 120
   heart attacks and strokes 121
   pulse rate 118–19
classification 74–5, 77
   plants 78–9
   using keys 75, 208–9
climate change 67, 176–7, 201
   changing the planet 58–9
climates 52
colds 41
communicating findings 132–3
communities 194
conclusions 201
condoms 149
conifers 78
conservation 60–1, 66, 195
   captive breeding 61
constipation 93
consumers 54, 184
correlations 119, 201
creative thinking 200
crops 176, 195, 217
Curiosity 29
cytoplasm 42

Darwin, Charles 218–19
   theory of evolution 220–1
decomposers 54
   recycling carbon 189, 197
   recycling minerals 188, 189, 197
   starting new food chains 188
deforestation 204, 207
depressants 146, 147
deserts 52–3, 174
diaphragm 124
diet 96–7, 121
dieticians 23

diffusion 112, 125
digestion 19, 157
digestive system 18, 19, 104–5, 110–11
diseases 40, 51, 149, 155
   defence against disease 150–1, 152–3
   insect-borne diseases 41, 113, 148, 150
dislocations 15
DNA 212
dormancy 168
dough 36, 50
drugs 146, 154–5
   illegal drugs 147
   pharmaceutical drugs 146
   social drugs 146

Earth's atmosphere 58–9
ecosystems 194
eggs 136, 137
ejaculation 136
embryos 136, 138–9
emulsification 107
endothermic reaction 158
energy 62–3, 67, 84
   food 94–5
   photosynthesis 158
energy flow 187
   passing on energy 186–7
   pyramids of biomass 187, 197
   pyramids of number 186
environment
   climate change 58–9, 176–7
   food chains 54–5
   food production 56–7
   habitats 52–3
   obtaining energy 62–3, 64–5
environmental variation 70–1, 211
enzymes 104, 106–7, 109
   food production 108–9
   how enzymes work 109
   providing right conditions 107
eutrophication 202
Everest 127
evidence 34–5
evolution 220, 227
   competition and selection 220
   inheritance of useful characteristics 221
   overproduction of offspring 220
   resistance 221
   survival of the fittest 220
excretion 28
exercise 118
exothermic reaction 158
extinction 60–1
   disrupted food webs 193
   invasive species 56, 192, 193
   long-term consequences 193
   lost species 192

faeces 40
fat cells 44
fats 92, 93, 96
fermentation 32, 36, 37, 51
   fermenters 33
fertilisation 136
   flowers 167
fertilisers 56
fetal alcohol syndrome 147
fetal development 138–9
fibre 93
filaments 166
fish 57, 175, 185, 202
fitness 120
flaccidity 88
flowering plants 8, 78
   growth 10–11, 12–13
   organs 8–9
   parasitic plants 9
flowers 8, 166, 171
   male and female parts 166
   pollination 166–7
flu 41
flue gas desulfurisation 199
food 92–3, 96–7, 102–3
   calculating how much you need 97
   choosing foods 100–1
   dietary deficiencies 98–9, 101
   testing for energy 94–5
food chains 54–5, 66, 82, 196
   energy flow 186–7
   starting new food chains 188
food poisoning 40
food production 56–7, 66
   climate change 176
   enzymes 108–9
food sterilisation 37
food webs 55, 184, 196–7
   disrupted food webs 193
   ocean food webs 185
   passing on energy 184–5
   trophic levels 185
Fossey, Dian 180
fossil fuels 62, 64
Fourier, Joseph 200
frequency charts 68
fruits 167
fuels 62–3, 64–5, 83
fungi 28, 31
   fungal diseases 40
Funk, Casimir 98

Galdikas, Biruté 180
Galen 20
gall bladder 107
Galvani, Luigi 46

gametes 166
gases
   acidic gases 198–9
   gas exchange 19, 125, 126–7
   greenhouse gases 59, 200
genes 42, 70, 210, 226
   dominant and recessive genes 214–15
   genes and chromosomes 212
   genetic engineering 222–3
genetic engineering
   adding genes 222
   copying genes 222
   growing medicines 222, 225
   pharming 223
   spiders' silk 223
   transferring genes to bacteria 223
geology 219
geothermal energy 63
germination 168
global warming 59, 200–1
glucose 19
gonorrhea 149
Goodall, Jane 180
graphs 118–19, 132
   curve of best fit 118
   line of best fit 119
greenhouse effect 59, 200
growth 28
   flowering plants 10–11, 12–13
   growth factors 49
   growth spurts 142
Guam 192–3
gut 104

habitats 52–3
   extreme habitats 174–5
   habitat destruction 56
   saving habitats 60–1
haematologists 23
haemoglobin 44
Hales, Stephen 11
hallucinogens 147
Harvey, William 20
health 121
heart 20, 116, 117
   heart attack 21, 121
   heart transplant 24
   tissues 48
hepatitis C 41
herbicides 56
herbivores 54
HIV 151
hormones 142
human body 26–7
   circulatory system 20–1
   medical science 22–5

Index

muscular system 16–17
organ systems 18–19
skeleton 14–15
hybrids 72
hydroponics 162
hyperaccumulators 164, 165
hyphae 31

Ibn al-Nafis 20
Ibn Sina 46
ideas 34–5
identifying trends 118–19
immune system 150–1, 152–3
implantation 138
inbreeding 195
infectious diseases 40, 51, 149, 155
infertility 72, 149
inherited features 70, 210, 226
   chromosomes 212–13
   dominant and recessive genes 214–15
   inheritance of useful characteristics 221
   Mendel, Gregor 214–15, 219
   selective breeding 216–17
insecticides 56
insects 75
   insect-borne diseases 41, 113, 148, 150
   pollination 166
insulators 175
interdependence 191
Intergovernmental Panel on Climate Change (IPCC) 201
intestines 19, 105
invasive species 56, 192
invertebrates 74–5, 80–1
   capturing small animals 179
investigations 200
iodine 159

joints 14–15, 82
   ball-and-socket joints 14
   hinge joints 14

keys to species identification 75, 208, 229
   designing keys 208–9
   using extra evidence 209
kidneys 112
   kidney transplant 24
kwashiorkor 99

lactic acid 32, 129
ladybirds 75
landslides 204
language 211
large intestine 105
leaves 8, 85
leishmaniasis 41
lichens 198
life 28, 50, 83
   identifying 29
   life on Mars 29
   seven characteristics 28
ligaments 15
Linnaeus, Carl 79
lipases 106, 108
Lister, Joseph 35
Loewi, Otto 47
lungs 124–5
   smoking 130–1
Lyell, Charles 219

magnification 30
malaria 41, 113
   resistant protozoa 224, 225
malnutrition 97
Malpighi, Marcello 11
mammals 76, 77, 81
mangrove forests 188–9
Mars 29
medical science 22–5
   artemisinin 224–5
   growing medicines 222
   pharming 223
Mendel, Gregor 214–15, 219
menopause 143
menstrual cycle 143
metal pollution 164, 165
micro-organisms 30, 34–5
   harmful micro-organisms 40–1
   use in food production 36–7
microbes 30–1
microscopes 30
   electron microscopes 30
Millennium Seed Bank, England 195
minerals 88, 92, 93, 162
   decomposers 188, 189, 197
mitochondria 128
movement 14, 28
   muscles 16–17
mucus 130
muscles 16, 18, 83
   anaerobic respiration 128–9
   antagonistic muscles 16
   control of muscle movement 17
   intercostal muscles 124
   muscle cells 44
   specialised muscles 129
   strength 17
mutations 217

naked mole rats 172
natural selection 219, 227
nerves 46
   experiments on frogs 46, 47
   sensing your surroundings 47

nervous system 18
neuroscientists 22
neutralisation 58
nicotine 131, 146
night blindness 98
non-renewable resources 62
nucleus 42
nutrition 28, 92–3, 156
nuts 167, 169

obesity 100
ocean floor 175
ocean food webs 185
optometrists 23
organic matter 33
organisms 28–9, 50–1, 83
organs 14, 18–19
    growing new organs 24
    scaffolds 25
    tissues 48
ovaries 136
    flowers 166
oviducts 136
ovulation 143
ovules 166
oxides of nitrogen 198
oxygen 85
    respiration 126, 127
ozone layer 58, 67

Pachauri, Rajendra 201
palm oil 64
pancreas 105
parasitic worms 148
Pasteur, Louis 32–3, 34–5
pasteurisation 33, 50
pathogens 150, 152
peak respiratory flow (PEF) 132
penis 136
periods 143
peristalsis 104
pH 107
phagocytes 150
pharming 223
phloem 45, 159
photobioreactors 65
photosynthesis 84–5, 158–9, 170
    preliminary tests 160–1
phytoextraction 164–5
placenta 138–9
plant cells 42–3, 82, 85
    cell wall 43
    specialised cells 45
plants 8–13, 28, 81, 82, 83, 162–3
    essential minerals 162
    estimating plant numbers 179, 183
    phytoextraction 164–5

plants without flowers 78–9
    saving water 174
    selective breeding 217, 224, 227
    survival 176–7
    testing growth 90–1, 156, 170–1, 228–9
    transpiration 88–9, 162
    why we need plants 84–5
    wilting 88
plasma 112
plasmids 223
platelets 112
pods 169
pollen 166, 167, 171
pollination 166–7, 214
pollution 56, 57, 64, 206–7
    air pollution 198–9
    metal pollution 164, 165
    water pollution 202–3
population 185, 197
    competition and selection 220
    overpopulation 57, 190–1
    predators and prey 191
    producers and consumers 190
predators and prey 54, 61, 172–3, 183, 191
predictions 34–5
preliminary tests 160–1
producers 54, 57
prosthetic limb developers 22
proteases 106
proteins 92, 96
protozoa 30, 41
    diseases 113
puberty 142, 143
pulse 118–19

quadrats 179

rainforest 52, 194
    making forests more valuable 204–5
    reducing demand for energy and
      resources 205
rectum 104, 105
renewable resources 62
reproduction 28, 144–5, 157
    budding 31
    eggs 136
    fertilisation 136
    making love 136–7
    overproduction of offspring 220
    plants 8
    sperm 136
reptiles 76
respiration 21, 28, 126–7, 134–5
    anaerobic respiration 128–9
    peak respiratory flow (PEF) 132
respiratory system 18, 19, 134, 135
    asthma 132–3

ribs 14
rice 57, 101
rickets 98
ringworm 40
river blindness 148
roots 8, 45

Sahara desert 174, 184
sampling techniques 178–9
  mark and recapture method 178
scavengers 54
scientific models 116
scientific questions 86
  choosing variables 86–7
  collecting results 87
  planning an investigation 86
  presenting results 87
scientists 22–3
  how scientists work 200–1
scurvy 98
seeds 167, 195
  germination 168
  seed dispersal 168–9, 171
selective breeding 216–17, 226–7
  mutations 217
  natural selection 219, 227
semen 136
sense organs 18
sensitivity 28
sewage 150, 202
sex hormones 142, 143
sexually transmitted infections 149
sickle cell anaemia 113
Sietas, Tom 126–7
skeleton 14–15, 18
  joints 14–15
skin transplants 24
skull 14
sleeping sickness 41, 113
small intestine 19, 105
smoking 130–1, 157
solar cells 62
solar energy 62
species 72–3
  invasive species 56, 192
  naming species 72–3, 79, 80
  species identification 208–9, 226, 229
specimens 30
sperm 136, 137
spinal cord 14
spongy mesophyll 85
spontaneous generation 34
spores 28, 78
sports scientists 22
sprains 15

stamens 166
starch 19, 85
stems 8
stigma 166
stimulants 146
stomach 104
stomata 85, 88, 89
streamlining 173
strokes 121
style 166
sugar 108
sulfur 58
sulfur dioxide 198, 199
survival 176–7
  survival of the fittest 220
synovial fluid 15
systems 18–19

tar 130
tendons 16
testes 136
theories 218
  evidence 219
  explanations 219
  making observations and asking questions 218–19
  theory of evolution 220–1
tissues 48
transplants 24
trees 10
trophic levels 185
tropical diseases 148
tuberculosis 149
turgidity 88
twins 140
  conjoined twins 141
  identical twins 71, 140, 210, 211, 229
  problems for twin babies 140–1
Tyndall, John 200
typhoid 40

ultraviolet radiation 58
urea 112
urine 112

vaccination 152–3
vacuoles 43, 165
vagina 136
van Helmont, Jan Baptista 10–11
variation 68–9, 80
  causes of variation 70–1, 210
  unique differences 69
vectors 41
veins 20, 116
vertebrae 14
vertebrates 74, 76–7, 80–1, 82

viruses 31, 41
vitamins 93, 98, 99

water 88–9, 162
    water energy 63
water pollution
    indicators 202
    monitoring rivers 202–3
    persistent organic pollutants 203
    sewage 202
wheat 101

wildlife sanctuaries 60
wind 167
    seed dispersal 169
    wind energy 63

xylem 45, 85, 88, 158

yeasts 31, 32–3
    bread 36
    working with yeasts 38–9
yoghurt 36, 50

OXFORD
UNIVERSITY PRESS

Great Clarendon Street, Oxford OX2 6DP

Oxford University Press is a department of the University of Oxford. It furthers the University's objective of excellence in research, scholarship, and education by publishing worldwide in

Oxford New York

Auckland Cape Town Dar es Salaam Hong Kong Karachi Kuala Lumpur Madrid Melbourne Mexico City Nairobi New Delhi Shanghai Taipei Toronto

With offices in

Argentina Austria Brazil Chile Czech Republic France Greece Guatemala Hungary Italy Japan Poland Portugal Singapore South Korea Switzerland Thailand Turkey Ukraine Vietnam

© Oxford University Press 2013

The moral rights of the authors have been asserted

Database right Oxford University Press (maker)

First published in 2013

All rights reserved. No part of this publication may be reproduced, stored in a retrieval system, or transmitted, in any form or by any means, without the prior permission in writing of Oxford University Press, or as expressly permitted by law, or under terms agreed with the appropriate reprographics rights organization. Enquiries concerning reproduction outside the scope of the above should be sent to the Rights Department, Oxford University Press, at the address above.

You must not circulate this book in any other binding or cover and you must impose this same condition on any acquirer

British Library Cataloguing in Publication Data

Data available

ISBN 978-0-19-839021-3

20 19 18

MIX Paper from responsible sources FSC® C007785

Printed in Great Britain by Bell and Bain Ltd, Glasgow

Acknowledgments

®IGCSE is the registered trademark of Cambridge International Examinations.

The publisher would like to thank Cambridge International Examinations for their kind permission to reproduce past paper questions.

Cambridge International Examinations bears no responsibility for the example answers to questions taken from its past question papers which are contained in this publication.

Cover photo: Eduardo Rivero / Shutterstock; **p8:** Cccsss/Dreamstime, Kajornyot/ Shutterstock, Trucic/Shutterstock; **p9:** Nasa; **p9:** Fletcher & Baylis/Science Photo Library; **p9:** Julia Pivovarova/Shutterstock; **p9:** Pudding/Bigstock; **p10:** Andreiuc88/Shutterstock; **p11:** Nigel Cattlin/Science Photo Library; **p14:** Perttu Sironen/Istock; **p14:** © Turhanerbas | Dreamstime.com; **p15:** Maxisport/Shutterstock.com; **p16:** Jeremyrichard / Shutterstock; **p16:** Maridav/Shutterstock; **p17:** Jacopin/Science Photo Library; **p17:** Yuriko Nakao/Reuters; **p18:** Sciepro/Science Photo Library; **p18:** Matthew Cole/ Shutterstock; **p18:** Jacopin/Science Photo Library; **p19:** Martin Dhorn/Royal College Of Surgeons/Science Photo Library; **p20:** Trip/Art Directors; **p20:** Sebastian Kaulitzki/ Shutterstock; **p21:** Nano/Istockphoto; **p22:** Peggy Greb/Us Department Of Agriculture/ Science Photo Library; **p22:** Colin Cuthbert/Science Photo Library; **p22:** James King-Holmes/Science Photo Library; **p23:** Life In View/Science Photo Library; **p23:** Adam Hart-Davis/Science Photo Library; **p24:** Alexander Tsiaras/Science Photo Library; **p24:** 3D4medical.com/Science Photo Library; **p24:** Aj Photo/Hop Americain/Science Photo Library; **p24:** Mauro Fermariello/Science Photo Library; **p25:** Sam Ogden/Science Photo Library; **p25:** Claudia Castillo/Associated Press; **p27:** 3D4medical.com/Science Photo Library; **p28:** Dr Keith Wheeler/Science Photo Library; **p28:** Prill Mediendesign & Fotografie / Istock; **p29:** Nasa; **p29:** Calvin J. Hamilton; **p29:** L. Newman & A. Flowers/ Science Photo Library; **p30:** Laguna Design/Science Photo Library; **p30:** Focal Helicopter/ Istockphoto; **p30:** Jacomstephens/Istockphoto; **p31:** David Scharf/Science Photo Library; **p31:** David Scharf/Science Photo Library; **p31:** Scimat/Science Photo Library; **p31:** Hazel Appleton, Centre For Infections/Health Protection Agency/Science Photo Library; **p32:** Jean-Loup Charmet/Science Photo Library; **p32:** Science Photo Library; **p33:** Maximilian Stock Ltd/Science Photo Library; **p33:** Paul Rapson/Science Photo Library; **p34:** Custom Medical Stock Photo/Science Photo Library; **p35:** Kings College School Of Medicine, Depart- Ment Of Surgery/Science Photo Library; **p36:** Idal/Shutterstock; **p36:** Power And Syred/Science Photo Library; **p37:** Tony Camacho/Science Photo Library; **p37:** Astrid & Hanns-Frieder Michler/Science Photo Library; **p38:** Sinclair Stammers/Science Photo Library; **p40:** Andy Crump/Science Photo Library; **p40:** Science Photo Library; **p40:** Cnri/ Science Photo Library; **p41:** Tim Vernon/Science Photo Library; **p41:** Russell Kightley/ Science Photo Library; **p42:** Dr Gopal Murti/Science Photo Library; **p42:** J.c. Revy, Ism/ Science Photo Library; **p44:** Professors P.m. Motta & S. Correr/Science Photo Library; **p44:** David Mack/Science Photo Library; **p44:** Prof. P. Motta/Dept. Of Anatomy/University "La Sapienza", Rome/Science Photo Library; **p45:** Steve Gschmeissner/Science Photo Library; **p45:** Steve Gschmeissner/Science Photo Library; **p46:** Science Source/Science Photo Library; **p46:** Sheila Terry/Science Photo Library; **p46:** Stephane Bidouze/ Shutterstock; **p46:** Leonello Calvetti/Shutterstock; **p47:** Andrea Danti/Shutterstock; **p50:** Georgy Markov/Shutterstock; **p52:** Stephane Bidouze/Shutterstock; **p52:** Som Chaij/ Shutterstock; **p52:** Tim Jenner/Shutterstock; **p52:** Stephen Dalton/Minden Pictures/Getty Images; **p53:** Nicolle Rager-Fuller, National Science Foundation/Science Photo Library; **p54:** Peter Ten Broecke/Istock; **p54:** Hedrus/Shutterstock; **p56:** Jeya Kumar/Istockphoto; **p56:** Volker Steger/Science Photo Library; **p58:** Anton Balazh/Shutterstock; **p58:** Nasa/ Science Photo Library; **p58:** Kenneth Keifer/Shutterstock; **p59:** Henning Dalhoff/Science Photo Library; **p60:** Mark Carwardine/Peter Arnold/Getty Images; **p60:** Carlos Caetano/ Shutterstock; **p61:** Dekanaryas/Shutterstock; **p61:** Sergei25/Shutterstock; **p62:** Tim Hughes/Istockphoto; **p61:** Jiri Foltyn/Shutterstock; **p61:** Prof. David Hall/Science Photo Library; **p63:** Cristovao/Shutterstock; **p63:** Pavle Marjanovic/Shutterstock; **p63:** N.minton/Shutterstock; **p64:** Tristan Tan/Shutterstock; **p64:** Vladimir Melnikov/ Shutterstock; **p65:** Pascal Goetgheluck/Science Photo Library; **p65:** Kurt G/Shutterstock; **p68:** William Perugini/Shutterstock; **p68:** Africa Studio/Shutterstock; **p69:** Erik Tham; **p70:** Konstantin Sutyagin/Shutterstock; **p70:** Goodgold99/Shutterstock; **p71:** Blend Images/Shutterstock; **p71:** Piti Tan & Stuart Jenner/Shutterstock; **p72:** Eric Isselee/ Shutterstock; **p72:** Martin Chow/Shutterstock; **p72:** Krzysztof Wiktor/Shutterstock; **p73:** Neal Grundy/Science Photo Library; **p74:** Gregory Dimijian/Science Photo Library; **p74:** Christopher Swann/Science Photo Library; **p76:** Anshu18/Shutterstock; **p76:** Joe Farah/ Shutterstock, Polushkin Ivan/Shutterstock; **p77:** Melissaf84/Shutterstock; **p77:** Dante Fenolio/Photo Researchers/Getty Images, Heiko Kiera/Shutterstock, Tom Mchugh/Science Photo Library; **p78:** Jim Steinberg/Science Photo Library; **p78:** Taina Sohlman/ Shutterstock; **p78:** Mimohe/Shutterstock; **p78:** Oksana2010/Shutterstock; **p79:** Becky Stares/Shutterstock; **p82:** Jgroup/Istockphoto; **p84:** Jim Barber/Shutterstock; **p85:** Power And Syred/Science Photo Library; **p85:** E. R. Degginger/Science Photo Library; **p86:** Suzanne Tucker/Shutterstock; **p88:** Taylor Hinton/Istock; **p89:** Dr Jeremy Burgess/ Science Photo Library; **p92:** Denis Kuvaev/Shutterstock; **p93:** Elena Schweitzer/ Shutterstock; **p94:** Erwinova/Shutterstock; **p96:** Maximilian Stock Ltd/Science Photo Library; **p96:** Iampuay/Shutterstock; **p96:** Shulevich/Fotolia; **p97:** Mariusz Prusaczyk/ Pantherstock; **p97:** Daniel Berehulak/Getty Images News/Getty Images, Juanmonino/ Istockphoto; **p98:** Biophoto Associates/Getty Images; **p98:** National Library Of Medicine; **p98:** Jeff Rotman/Peter Arnold/Getty Images; **p99:** Andy Crump, Tdr, Who/Science Photo Library; **p99:** Paul Cowan/Shutterstock; **p100:** Tarasov/Shutterstock; **p100:** Luis Louro/ Shutterstock; **p100:** Carlos Dominguez/Science Photo Library; **p101:** Ton Kinsbergen/ Science Photo Library; **p101:** International Rice Research Institute (Irri); **p103:** Biophoto Associates/Science Photo Library; **p104:** Suhendri/Bigstock; **p108:** Altrendo Images/ Altrendo/Getty Images; **p112:** Herve Conge, Ism/Science Photo Library; **p112:** Eye Of Science/Science Photo Library; **p113:** Eye Of Science/Science Photo Library; **p113:** Dr M.a. Ansary/Science Photo Library; **p113:** Daniel Snyder/Visuals Unlimited/Getty Images; **p114:** Ariel Skelley/Blend Images/Getty Images; **p114:** Cnri/Science Photo Library, Ed Reschke/ Peter Arnold/ Getty Images; **p115:** Asem Arab/Fotolia; **p116:** Bokica I/Bigstock; **p118:** Shaun Botterill - Fifa/Getty Images; **p120:** Apurv Goyal/Q2a Image Bank; **p120:** Bsip Vem/Science Photo Library; **p121:** P. Marseaud, Ism/Science Photo Library; **p124:** Hybrid Medical Animation / Science Photo Library; **p124:** Martin Dohrn/Royal College Of Surgeons/ Science Photo Library; **p125:** Dr. Gladden Willis, Visuals Unlimited /Science Photo Library; **p126:** Richard Drew/Associated Press; **p127:** Zzvet/Dreamstime.com; **p128:** Giuliano Bevilacqua/Rex Features; **p128:** Medimage/Science Photo Library; **p130:** Matt Meadows/Getty Images; **p130:** Juergen Berger/Science Photo Library; **p130:** Du Cane Medical Imaging Ltd/Science Photo Library; **p131:** Hkannn/Shutterstock; **p132:** B. Boissonnet/Bsip/Glow Images; **p133:** Gary Carlson/Science Photo Library; **p133:** Lusoimages; **p136:** Samuel Borges/Bigstock; **p136:** Eye Of Science/Science Photo Library; **p138:** Dr G. Moscoso/Science Photo Library; **p139:** Centers For Disease And Control Prevention (Cdc); **p140:** Lwa/Dann Tardif/Blend Rf/Glow Images; **p141:** Reflekta Studios/ Fotolia; **p141:** Luis Enrique Ascui/Stringer/Getty Images News/Getty Images; **p142:** Apollofoto/Bigstock; **p146:** Ansar80/Shutterstock; **p146:** Dmitry Kalinovsky/ Shutterstock; **p146:** Science Photo Library; **p147:** Dorling Kindersley/Getty Images; **p147:** Susan Astley/University Of Washington; **p148:** A. Crump, Tdr, Who/Science Photo Library; **p148:** Herve Conge, Ism /Science Photo Library And Dr. Shirley Maddison/ Centers For Disease And Control Prevention (Cdc); **p148:** A. Crump, Tdr, Who/Science Photo Library; **p148:** Mark Giles/Science Photo Library; **p148:** Centers For Disease And Control Prevention (Cdc); **p149:** Denis Meyer/Imagebroker/Glow Images; **p149:** Bsip/ Universal Images Group/Getty Images; **p149:** Th Foto-Werbung/Science Photo Library; **p149:** Western Ophthalmic Hospital/Science Photo Library; **p150:** Tony Karumba/ Stringer/Afp/Getty Images; **p150:** Centers For Disease And Control Prevention (Cdc); **p150:** Science Photo Library; **p151:** Sebastian Kaulitzki/Shutterstock; **p151:** Nibsc/ Science Photo Library; **p152:** Hazel Appleton, Centre For Infections/Health Protection Agency/Science Photo Library; **p152:** Zurijeta/Shutterstock; **p153:** Centers For Disease And Control Prevention (Cdc); **p153:** Ggw1962/Shutterstock; **p158:** Treasure Dragon/ Shutterstock; **p158:** Heitipaves/Dreamstime.com; **p159:** Nigel Cattlin/Flpa; **p159:** Dvarg/ Shutterstock; **p162:** Protasov An/Shutterstock; **p163:** Sura Nualpradid/Shutterstock; **p164:** Ria Novosti/Science Photo Library; **p164:** Art Konovalov/Shutterstock; **p165:** Forest & Kim Starr; **p166:** Blueringmedia/Shutterstock; **p166:** Blueringmedia/ Shutterstock; **p166:** Mr. Green/Shutterstock; **p168:** Bogdan Wankowicz/Shutterstock; **p168:** Anest/Shutterstock; **p172:** Aughty/Dreamstime.com; **p172:** Tania Thomson/ Shutterstock; **p172:** Krzysztof Wiktor/Shutterstock; **p172:** Mitch Reardon/Science Photo Library; **p174:** Patrick Poendl/Shutterstock; **p174:** Pichugin Dmitry/Shutterstock; **p174:** Annettt/Shutterstock; **p175:** Hellio & Van Ingen/Nhpa/Photoshot; **p175:** Dante Fenolio/ Science Photo Library; **p176:** Galyna Andrushko/Shutterstock; **p176:** Sergey Uryadnikov/ Shutterstock; **p177:** William Ervin/Science Photo Library; **p178:** Villiers Steyn/ Shuttersock; **p178:** Niall Dunne/Shutterstock; **p179:** Martyn F. Chillmaid/Science Photo Library; **p180:** Jane Goodall Institute; **p180:** Andy Gehrig/Istock; **p181:** Albie Venter/ Shutterstock; **p181:** Rich Carey/Shutterstock; **p184:** Lenar Musin/Shutterstock; **p186:** Johan Barnard/Shutterstock; **p188:** Vilainecrevette/Shutterstock; **p189:** Nasa; **p190:** Steve Estvanik/Shutterstock; **p191:** Nancys/Shutterstock; **p191:** Scott E Read/ Shutterstock; **p192:** John Mitchell/Science Photo Library; **p192:** Anthony Mercieca/ Science Photo Library; **p193:** Fotomak/Shutterstock; **p193:** Szefei/Shutterstock; **p194:** Pichugin Dmitry/Shutterstock; **p194:** Frontpage/Shutterstock; **p195:** James King-Holmes/ Science Photo Library; **p195:** Albie Venter/Shutterstock; **p198:** Suzz/Shutterstock; **p200:** Forrest M. Mims Iii; **p201:** Joze Pojbic/Istockphoto; **p202:** Evlakhov Valeriy/Shutterstock; **p202:** John Kasawa/Shutterstock; **p204:** Hagit Berkovich/Shutterstock; **p204:** Prill/ Shutterstock; **p204:** Poznyakov/Shutterstock; **p208:** Dennis Donohue/Shutterstock; **p210:** Herjua/Shutterstock; **p211:** Snowwhiteimages/Shutterstock; **p211:** Snowwhiteimages/Shutterstock; **p212:** Dimarion/Shutterstock; **p216:** Eric Isselee/ Shutterstock; **p216:** Pavel Sazonov /Bigstock; Andi Pantz /Istockphoto; Michael Chen/ Istockphoto; **p216:** Jl Creative Captures; **p217:** Smereka/Shutterstock; **p217:** Pinthong Nakon/Shutterstock; **p218:** Tischenko Irina/Shutterstock; **p218:** Morphart Creation/ Shutterstock; **p218:** Eric Isselee/Shutterstock; **p219:** 88136284/Shutterstock; **p219:** Catwalker/Shutterstock; **p220:** Kitch Bain/Shutterstock; **p220:** Dennis Donohue/ Shutterstock; **p221:** Oleg Kozlov/Shutterstock; **p222:** Eye Of Science/Science Photo Library; **p223:** Philippe Psaila/Science Photo Library; **p224:** Andy Crump, Tdr, Who/ Science Photo Library; **p224:** Scott Bauer/Us Department Of Agriculture/Science Photo Library; **p225:** Hank Morgan/Science Photo Library;

Artwork by: Q2A Media, Erwin Haya, Barking Dog Art and OUP.